Markov Random Fields

Yu. A. Rozanov

Markov Random Fields

Translated by Constance M. Elson

Springer-Verlag
New York Heidelberg Berlin

Yu. A. Rozanov
Steklov Mathematics Institute
UL. Vavilov 42
Moscow 117333
U.S.S.R.

Constance M. Elson (*Translator*)
403 Turner Place
Ithaca, NY 14850
U.S.A.

AMS Subject Classifications (1980): 60G60

Library of Congress Cataloging in Publication Data

Rozanov, ĪU. A. (ĪUriĭ Anatol'evich), 1934–
    Markov random fields.

    (Applications of mathematics)
    Bibliography: p.
    Includes index.
    1. Random fields.   2. Vector fields.
I. Title.   II. Series.
QA274.45.R68   519.2   82-3303
                                 AACR2

With 1 Illustration

The original Russian title is Random Vector Fields, published by Nauka 1980.

Typeset by Composition House Ltd., Salisbury, England.

9 8 7 6 5 4 3 2 1

ISBN-13:978-1-4613-8192-1   e-ISBN-13:978-1-4613-8190-7
DOI: 10.1007/978-1-4613-8190-7

Preface

In this book we study Markov random functions of several variables.

What is traditionally meant by the Markov property for a random process (a random function of one time variable) is connected to the concept of the phase state of the process and refers to the independence of the behavior of the process in the future from its behavior in the past, given knowledge of its state at the present moment. Extension to a generalized random process immediately raises nontrivial questions about the definition of a suitable "phase state," so that given the state, future behavior does not depend on past behavior. Attempts to translate the Markov property to random functions of multi-dimensional "time," where the role of "past" and "future" are taken by arbitrary complementary regions in an appropriate multi-dimensional time domain have, until comparatively recently, been carried out only in the framework of isolated examples. How the Markov property should be formulated for generalized random functions of several variables is the principal question in this book. We think that it has been substantially answered by recent results establishing the Markov property for a whole collection of different classes of random functions. These results are interesting for their applications as well as for the theory. In establishing them, we found it useful to introduce a general probability model which we have called a random field.

In this book we investigate random fields on continuous time domains.

Contents

CHAPTER 3

The Markov Property for Generalized Random Functions 103

CHAPTER 1

General Facts About Probability Distributions

§1. Probability Spaces

1. Measurable Spaces

Let X be an arbitrary set. When we consider elements $x \in X$ and sets $A \subseteq X$, we call X a *space.*

We use standard notation for set operations: $\cup$ for union, $\cap$ for intersection (also called the product and sometimes indicated by a dot), A^c for the complement of A, $A_1 \backslash A_2 = A_1 \cdot A_2^c$ for the difference of A_1 and A_2, $A_1 \circ A_2 = (A_1 \backslash A_2) \cup (A_2 \backslash A_1)$ for the symmetric difference, $\varnothing$ for the empty set.

Collections of Sets. When looking at collections of sets, we will use the following terminology.

A collection $\mathfrak{G}$ of subsets of the space X is called a *semi-ring* when for any sets A, A_1 in $\mathfrak{G}$ their intersection is also in $\mathfrak{G}$ and when $A_1 \subseteq A$, then A can be represented as a finite union of disjoint sets $A_1, \ldots, A_n$ in $\mathfrak{G}$, $A = \bigcup_1^n A_i$. We also require that $\varnothing \in \mathfrak{G}$ and the space X itself be represented as a countable union of disjoint sets $A_1, \ldots \in \mathfrak{G}$: $X = \bigcup_1^\infty A_i$.

A semi-ring $\mathfrak{G}$ is a *ring* if for any two sets A_1, A_2, it also contains their union.

Let $\mathfrak{G}$ be an arbitrary semi-ring. Then the collection of all sets $A \subseteq X$ which can be represented as a finite union of intersections of sets in $\mathfrak{G}$ is a ring. If the ring $\mathfrak{G}$ also includes the set X, then it is called an *algebra.*

An algebra is invariant with respect to the operations union, intersection and complement, taken a finite number of times. The collection of sets is called a *σ-algebra* if this invariance holds when the operations are taken a countable number of times.

The intersection of an arbitrary number of σ-algebras is again a σ-algebra. For any collection of sets $\mathfrak{G}$, there is a σ-algebra $\mathscr{A}$ containing $\mathfrak{G}$. The minimal such σ-algebra is called the *σ-algebra generated by the collection* $\mathfrak{G}$.

EXAMPLE (Union of σ-algebras). Let $\mathscr{A} = \mathscr{A}_1 \vee \mathscr{A}_2$ be the minimal σ-algebra containing both $\mathscr{A}_1$ and $\mathscr{A}_2$. It is generated by the semi-ring $\mathfrak{G} = \mathscr{A}_1 \cdot \mathscr{A}_2$ of sets of the form $A = A_1 \cdot A_2$, $A_i \in \mathscr{A}_i$.

We call a σ-algebra $\mathscr{A}$ *separable* if it is generated by some countable collection of sets $\mathfrak{G}$. Notice that in the case when $\mathfrak{G}$ is a countable collection, the algebra it generates is countable, consisting of all sets which can be derived from $\mathfrak{G}$ by finite intersections, unions, and complements.

When we speak of X as a *measurable space* we will mean that it is equipped with a particular σ-algebra $\mathscr{A}$ of sets $A \subseteq X$. We indicate a measurable space by the pair $(X, \mathscr{A})$. In the case where X is a topological space, then frequently the σ-algebra $\mathscr{A}$ is generated by a complete neighborhood system (basis) of X. Usually we will deal with the *Borel σ-algebra*, generated by all open (closed) sets, or the *Baire σ-algebra*, which is the σ-algebra generated by inverse images of open (closed) sets in $\mathbb{R}$ under continuous mappings $\varphi \colon X \to \mathbb{R}$.

If X is a metric space with metric ρ, and if $F \subseteq X$ is any closed set, then the function $\varphi(x) = \inf_{x' \in F} (x, x')$, $x \in X$, is continuous and F is the preimage of $\{0\}$ under φ, $F = \{x \colon \varphi(x) = 0\}$; hence each Borel set is Baire. This is also true for compact X with countable basis: such a space is metrizable.

EXAMPLE. The system of half-open intervals (x', x'') on the real line $X = \mathbb{R}$ forms a semi-ring and the σ-algebra it generates is the collection of all Borel sets. The same is true of the countable semi-ring of half-open intervals with rational endpoints.

EXAMPLE (The semi-ring generated by closed sets). The collection $\mathfrak{G}$ of all sets of the form $A = G_1 \backslash G_2$, where G_1 and G_2 are closed sets, is a semi-ring: for any A', $A'' \in \mathfrak{G}$, $A' \cap A'' = G_1' \cdot G_1'' \backslash (G_2' \cup G_2'') \in \mathfrak{G}$; furthermore if $A'' \subseteq A'$, we can assume $G_2' \subseteq G_2'' \subseteq G_1'' \subseteq G_1'$ and we have $A'\backslash A'' = A_1 \cup A_2$ where $A_1 = G_1'\backslash G_1''$ and $A_2 = G_2''\backslash G_2'$ are disjoint.

EXAMPLE (The semi-ring of Baire sets). Let F be a closed Baire set in X which is the inverse image of some closed set B on the real line Y, $F = \{\varphi \in B\}$. If one takes any continuous function ψ on Y, mapping the closed set B to 0 and strictly positive outside B (for instance, $\psi(y)$ could be the distance from the point $y \in Y$ to the set $B \subseteq Y$), then the composition $\psi \circ \varphi$ is continuous on X and the closed Baire set F is precisely the null set $\{\psi \circ \varphi = 0\}$. The system $\mathfrak{G}$ of all closed Baire sets F which are null sets of continuous functions φ on the real line contains the intersection $F_1 \cap F_2$ and union $F_1 \cup F_2$ for any F_1, $F_2 \in \mathfrak{G}$. For example, if $F_i = \{\varphi_i = 0\}$ then $F_1 \cup F_2 =$

$\{\varphi_1\varphi_2 = 0\}$ and $F_1 \cap F_2 = \{|\varphi_1| + |\varphi_2| = 0\}$. The collection of all sets A which can be represented as a difference $F_1 \backslash F_2$ of two sets $F_2 \subseteq F_1$ in $\mathfrak{G}$ is a semi-ring which generates the entire σ-algebra of Baire sets in the space X.

Standard Borel σ-algebras. Let $(X, \mathscr{A})$ be a measurable space; we call $\mathscr{A}$ a standard Borel σ-algebra if it is isomorphic to a Borel σ-algebra $\mathscr{B}$ on some Borel subset Y of a complete separable metric space. (Two σ-algebras $\mathscr{A}$ and $\mathscr{B}$ are Borel isomorphic if there is a one-to-one mapping $\varphi\colon X \to Y$ and $\mathscr{A}$ consists of all $A \subseteq X$ of the form $A = \{x\colon \varphi(x) \in B\}, B \in \mathscr{B}$.) The following holds: *a standard Borel σ-algebra is isomorphic to a Borel σ-algebra on some compact metric space.*

Products of Spaces. The product of measurable spaces $(X_1, \mathscr{A}_1)$ *and* $(X_2, \mathscr{A}_2)$ is the space $X = X_1 \times X_2$ of all pairs (x_1, x_2), $x_i \in X_i$, with σ-algebra $\mathscr{A}$ generated by the semi-ring $\mathfrak{G} = \mathscr{A}_1 \times \mathscr{A}_2$ of sets $A \subseteq X$ of the form $A = A_1 \times A_2, A_i \in \mathscr{A}_i$; more precisely, A is the set of all pairs (x_1, x_2), $x_i \in A_i$.

We define a finite product $X = \prod_{t \in T} X_t$ of measurable spaces $(X_t, \mathscr{A}_t)$ in the same way. Here T is a finite index set and X is the set of elements $x = \{x_t, t \in T\}$, each a tuple of "coordinates" $x_t \in X_t$, with σ-algebra $\mathscr{A}$ generated by the semi-ring $\mathfrak{G} = \prod_{t \in T} \mathscr{A}_t$ of sets $A = \prod_{t \in T} A_t$, $A_t \in \mathscr{A}_t$. Each such A is a set of elements x with corresponding coordinates $x_t \in A_t$.

Let T be an arbitrary index set and $(X_t, \mathscr{A}_t), t \in T$, be an arbitrary family of measurable spaces. We define the product $X = \prod_{t \in T} X_t$ to be the space of elements $x = \{x_t, t \in T\}$, given by means of "coordinates" $x_t \in X_t$, with σ-algebra $\mathscr{A}$ generated by the semi-ring $\mathfrak{G} = \prod_{t \in T} \mathscr{A}_t$ of cylinder sets. A *cylinder set* $A \subseteq X$ is of the form

$$A = \{x\colon x_S \in A_S\}, \tag{1.1}$$

where S is a finite subset of T. Here the symbol x_S indicates the point in the space $X_S = \prod_{t \in S} X_t$ whose S-coordinates are the same as those of x and $A_S \subseteq X_S$ is a set in the semi-ring $\prod_{t \in S} \mathscr{A}_t$. We call $(X, \mathscr{A})$ a coordinate space.

If the X_t are topological spaces, then the cylinder sets (1.1) with A_S of the form $A_S = \prod_S A_t$, A_t open in X_t, form a basis for the topological space $X = \prod_T X_t$; this is the *Tychonov product.* A commonly used example is the coordinate space $X = E^T$; here each X_t is some fixed space E and $\mathscr{A}_t$ is some fixed σ-algebra $\mathscr{B}$. The elements $x = \{x(t), t \in T\}$, of this space are all possible functions on the set T with values in the "phase space" E.

2. Distributions and Measures

A non-negative function $P = P(A)$ defined on the semi-ring $\mathfrak{G}$ of sets A in the space X is a *distribution* if $P(\varphi) = 0$ and

$$P(A) = \sum_k P(A_k), \quad \text{whenever } A = \bigcup_k A_k, \text{ a countable union of disjoint}$$

$$\text{sets } A_1, \ldots, \text{ in } \mathfrak{G}. \tag{1.2}$$

In case (1.2) is true only for a finite number of sets, the function P is usually called a *weak distribution*. Every weak distribution P can be uniquely extended from the semi-ring $\mathfrak{G}$ to the ring of all sets $A \subseteq X$ which are a finite union of disjoint sets $A_1, \ldots, A_n \in \mathfrak{G}$; the extension is done using (1.2), which gives the finite additivity. A (weak) distribution P on a semi-ring $\mathfrak{G}$ is called bounded if the function $P(A)$ is bounded. We consider only bounded distribution. A weak distribution P on a ring $\mathfrak{G}$ is a distribution iff it is continuous in the following sense: for every monotone sequence of sets $A_1 \supseteq A_2 \supseteq \cdots$ whose intersection $\bigcap_n A_n = \varnothing$, $\lim_{n \to \infty} P(A_n) = 0$.

Each distribution extends uniquely to a measure, i.e., a countably additive function P on the σ-algebra $\mathscr{A}$ generated by the original semi-ring $\mathfrak{G}$. The extension is defined by

$$P(A) = \inf \sum_k P(A_k), \tag{1.3}$$

where the inf is taken over all sets $A_1, A_2, \ldots \in \mathfrak{G}$ whose union contains the set A.

A measure P on a topological space X is *Borel (Baire)* if it is defined on the Borel (Baire) sets.

For any set $A \subseteq X$, define $P(A)$ by means of (1.3); for $A_1, A_2 \subseteq X$, the "distance" $\rho(A_1, A_2) = P(A_1 \circ A_2)$ indicates to what extent the sets A_1, A_2 differ from one another. Let P be a measure on the σ-algebra $\mathscr{A}$. A set $A \subseteq X$ is called *measurable* if $\exists$ some $A' \in \mathscr{A}$ such that $P(A \circ A') = 0$. If $\mathfrak{G}$ is a ring generating $\mathscr{A}$, then a set A is measurable iff it can be approximated by sets $A_\varepsilon \in \mathfrak{G}$ in the sense that

$$P(A \circ A_\varepsilon) \leq \varepsilon, \qquad \text{for any } \varepsilon > 0. \tag{1.4}$$

The collection of all measurable sets is a σ-algebra and (1.3) defines the measure P on it. This extension of the original measure P is *complete* in the sense that any subset A' of a set A of measure zero is measurable and $P(A') = 0$.

All of the above observations apply to unbounded distributions and measures with minor restrictions; in discussing unbounded measures it is important to stress that X must be σ-finite, i.e., representable as a countable union of sets of finite measure.

Let $\mathscr{A}_1, \mathscr{A}_2$ be two collections of sets having the property that for each $A_1 \in \mathscr{A}_1$ and $A_2 \in \mathscr{A}_2$, one can find $A_1' \in \mathscr{A}_2$ and $A_2' \in \mathscr{A}_1$ differing from A_1, A_2 by sets of measure 0, $P(A_1 \circ A_1') = P(A_2 \circ A_2') = 0$. We indicate this situation by the equality

$$\mathscr{A}_1 = \mathscr{A}_2 \ (\text{mod } 0).$$

Tight Measures. A Borel measure P on a topological space X is *regular* if for every measurable set A,

$$P(A) = \sup_{F \subseteq A} P(F), \tag{1.5}$$

where the sup is taken over all closed sets $F \subseteq A$. (1.5) is equivalent to

$$P(A) = \inf_{G \supseteq A} P(G), \tag{1.6}$$

where the inf is taken over all open G containing A.

Let the measure P have the property that $P(X) = \sup_{F \subseteq X} P(F)$, where the sup is taken over compact F. Such P is said to be *tight*. Every measure on a complete separable metric space is tight. For such measures (1.5) can be restated with "compact" replacing "closed": i.e., a regular tight measure is *Radon*.

EXAMPLE Let X be the real line. Then the Borel and Baire sets coincide. The measure $P(A)$ of each measurable set A is defined by (1.3), where the inf is taken over all disjoint half-open intervals $A_k = (x'_k, x''_k]$ whose union contains A. At the same time each interval $(x', x'']$ is the intersection of a countable number of open intervals and (1.6) clearly holds with the inf taken over all open sets G containing A.

Equation (1.6) is also true in any topological space X on which the Borel and Baire sets coincide. Every set $F \in \mathscr{A}$ which is the null set of a continuous function φ on the real line Y is the intersection of a countable number of open sets of the form $G_\delta = \{|\varphi| < \delta\}$, $F = \bigcap G_\delta$. By the continuity of the measure P, $P(F) = \inf P(G_\delta)$. For the difference of such sets, $A = F_1 \backslash F_2$ with $F_2 \subseteq F_1$, we have $P(A) = P(F_1) - P(F_2) = \inf P(G)$, with the inf taken over all open sets G of the form $G = G_1 \backslash F_2$, with $F_1 \subseteq G_1$. Since sets of the form $F_1 \backslash F_2$ are a semi-ring generating $\mathscr{A}$, we have for any P-measurable set A, $P(A) = \inf \sum_k P(A_k)$, where the A_k are a countable disjoint covering of A and each $A_k = F_{1k} \backslash F_{2k}$. Clearly $P(A)$ coincides with $\inf_{G \supseteq A} P(G)$, the inf taken over all unions G of appropriate sets.

A weak distribution P on a semi-ring $\mathfrak{G}$ is *tight* if each set $A \in \mathfrak{G}$ can be arbitrarily closely approximated in the sense (1.4) by compact sets $F_\varepsilon \subseteq A$: $P(A \backslash F_\varepsilon) \le \varepsilon$ for any $\varepsilon > 0$. Such a weak distribution is a distribution and extends to a tight measure on the σ-algebra generated by $\mathfrak{G}$.

We will show why this is true. We assume $\mathfrak{G}$ is the ring formed by finite unions of sets in the original semi-ring. It is sufficient to establish that P is continuous. If $\lim_n P(A_n) \ne 0$ for some sequence $A_1 \supseteq A_2 \supseteq \cdots$, then one can find a sequence of approximating compacts $F_n \subseteq A_n$ with $F_1 \supseteq F_2 \supseteq \cdots$, and with $P(F_n) = P(A_n) - P(A_n \backslash F_n) > 0$ and whose intersection is nonempty, $\varnothing \ne \bigcap_n F_n \subseteq \bigcap_n A_n$. Hence for any sequence $A_1 \supseteq A_2 \supseteq \cdots$ whose intersection is empty, we have $\lim P(A_n) = 0$.

Products of Measures. Let P_1, P_2 be measures on measurable spaces $(X_i, \mathscr{A}_i)$, $i = 1, 2$. The equation

$$P(A) = P_1(A_1) P_2(A_2)$$

defines a distribution on the semi-ring $\mathfrak{G} = \mathscr{A}_1 \times \mathscr{A}_2$ (sets of the form $A = A_1 \times A_2$, $A_i \in \mathscr{A}_i$) in the product space $X = X_1 \times X_2$. The corresponding measure $P = P_1 \times P_2$ on the σ-algebra $\mathscr{A}$ generated by $\mathscr{A}_1 \times \mathscr{A}_2$ is the *product* of the *measures* P_1 and P_2.

For an arbitrary family of measure spaces $(X_t, \mathscr{A}_t)$, $t \in T$, with $P_t(X_t) = 1$ for all but a finite number of t, we define the *product measure* in a similar way: $P = \prod_{t \in T} P_t$ on the coordinate space $X = \prod_{t \in T} X_t$ with σ-algebra $\mathscr{A}$ generated by the semi-ring $\mathfrak{G} = \prod_{t \in T} \mathscr{A}_t$ of cylinder sets of the form (1.1).

Let $X = \prod_{t \in T} X_t$ be a coordinate space and P a distribution on the semiring $\mathfrak{G} = \prod_T \mathscr{A}_t$ of cylinder sets (1.1). Then

$$P_S(A_S) = P(A), \qquad A \in \mathfrak{G} \tag{1.7}$$

defines the projection of the distribution P on the space $X_S = \prod_{t \in S} X_T$ and the corresponding semi-ring $\prod_{t \in S} \mathscr{A}_t$. It satisfies the following consistency condition: for $S_1 \subseteq S_2$, the distribution P_{S_1} is the projection of the distribution P_{S_2}.

Let P_S, $S \subseteq T$, be a family of distributions parametrized by finite subsets $S \subseteq T$ and satisfying the consistency conditions described above. Then equation (1.7) defines a weak distribution $P = P_T$ on the space $X = \prod_{t \in T} X_t$ and the semi-ring $\prod_{t \in T} \mathscr{A}_t$. For an arbitrary $S \subseteq T$, let P_S denote the projection of P on the space X_S and semi-ring $\prod_{t \in S} \mathscr{A}_t$. Clearly P_T is a distribution $\Leftrightarrow$ for any countable $S \subseteq T$, P_S is a distribution since then equation (1.2) will hold for countably many cylinder sets in $\prod_{t \in T} \mathscr{A}_t$.

In the case of a topological space, we saw that a weak distribution P_S is a distribution if it is tight. Suppose that the distributions P_t corresponding to singleton sets $S = \{t\}$, are tight. Then $P = P_S$ will be tight for countable $S \subseteq T$ since each set $A = \prod_{t \in S} A_t$, $A_t \in \mathscr{A}_t$, can be approximated arbitrarily closely by compacts of the form $F = \prod_{t \in S} F_t = \bigcap_{t \in S} \{x_t \in F_t\}$ for a suitable choice of $F_t \subseteq X_t$:

$$A \backslash F = \bigcup_{t \in S} \{x_t \in A_t \backslash F_t\}, \qquad P(A \backslash F) \leq \sum_{t \in S} P_t(A_t \backslash F_t).$$

In particular, if $X = E^T$, where E is a complete separable metric space, then for a consistent family of distributions P_S corresponding to finite $S \subseteq T$, equation (1.7) gives a distribution P on cylinder sets and it can be extended to a measure on the σ-algebra generated by the semi-ring $\mathfrak{G} = \prod_{t \in T} \mathscr{A}_t$.

Let $X = \prod_{t \in T} X_t$ be an arbitrary coordinate space with measure P on the σ-algebra $\mathscr{A}(T)$ generated by the semi-ring $\prod_{t \in T} \mathscr{A}_t$. Then for each measurable $A \subseteq X$, $\exists$ some countable $S \subseteq T$ and set A' in the σ-algebra $\mathscr{A}(S)$ generated by the semi-ring $\prod_{t \in S} \mathscr{A}_t$ such that $A = A'$ (mod 0); that is, A and A' differ by a set of measure 0.

Mappings and Measures. Let $(X, \mathscr{A})$ be a measurable space with measure P on the σ-algebra $\mathscr{A}$ and let $\varphi(x)$, $x \in X$, be a function taking values in a space Y. The equation

$$P^{\varphi}(B) = P\{\varphi \in B\} \tag{1.8}$$

defines a measure P^φ on the σ-algebra $\mathscr{B}$ consisting of all sets $B \subseteq \mathscr{B}$ whose pre-images $A = \{\varphi \in B\}$ belong to the σ-algebra $\mathscr{A}$.

Let X be an arbitrary space and $(Y, \mathscr{B})$ a measurable space with measure Q on the σ-algebra $\mathscr{B}$. We write $\varphi(A)$ for the image of a set $A \subseteq X$ under the map $\varphi \colon X \to Y$. When the set $\varphi(X)$ is measurable and $A = \{\varphi \in B\}$, then

$$P(A) = Q(B \cap \varphi(X)) \tag{1.9}$$

defines a measure P on the σ-algebra $\mathscr{A}^\varphi$, consisting of all pre-images $A = \{\varphi \in B\}$, $B \in \mathscr{B}$; we will say that the σ-algebra $\mathscr{A}^\varphi$ is generated by the function φ.

A map φ from a measurable space $(X, \mathscr{A})$ with complete measure P on the σ-algebra $\mathscr{A}$ to the measurable space $(Y, \mathscr{B})$ is called measurable if for each $B \in \mathscr{B}$, the pre-image $A = \{\varphi \in B\}$ is a measurable set. When speaking of a real measurable function φ, we will mean a map to the real line $Y = \mathbb{R}$ with the Borel σ-algebra $\mathscr{B}$.

Let $(X, \mathscr{A})$ be a topological space with tight Borel measure P. Then for any real measurable function φ, the image $B = \varphi(X)$ is a measurable set of the real line (with respect to the corresponding Borel measure P^φ).

We will show this. A measurable function φ is the uniform limit of piecewise constant functions φ_n defined by $\varphi_n(x) = y_{kn}$ if $x \in A_{kn}$, where $A_{1n}, \ldots$ are disjoint measurable sets in X; one can take approximating compact sets $F_{kn} \subseteq A_{kn}$ whose finite unions $F_n = \bigcup_{k \le m_k} F_{kn}$ are such that the intersection $X_\varepsilon = \bigcap_n F_n$ approximates the space X to within any previously specified $\varepsilon > 0$:

$$P(X \backslash X_\varepsilon) < \varepsilon.$$

Each function φ_n on the compact set X_ε takes only a finite number of different values y_{kn}; moreover, the pre-images $\{\varphi_n = y_{kn}\} = F_{kn} \cap X_\varepsilon$ are compact. It is clear that all functions $\varphi_n(x)$, $x \in X_\varepsilon$, as well as their uniform limit $\varphi(x)$, are continuous on X_ε. Let $B = \varphi(X)$. The image $B_\varepsilon = \varphi(X_\varepsilon)$ is compact, since φ is continuous, and we have

$$P^\varphi(B \backslash B_\varepsilon) = P(\{\varphi \in B\} - \{\varphi \in B_\varepsilon\}) \le P(X \backslash X_\varepsilon) < \varepsilon,$$

where ε can be chosen to be arbitrarily small. By (1.4) the set B is measurable.

A measure P on a measurable space (X, A) will be called perfect if for each measurable real function φ on X the image $B = \varphi(X)$ is measurable with respect to the Borel measure P^φ.

The Weak Topology on the Space of Measures. Let X be a topological space and $\mathscr{M}(X)$ the collection of all Borel measures P on the space X, normalized so $P(X) = 1$. The *weak topology* on $\mathscr{M}(X)$ is the topology generated by neighborhoods of $P \in \mathscr{M}(X)$ of the type

$$\left\{ \tilde{P} \colon \left| \int_X f \, d\tilde{P} - \int_X f \, dP \right| < \varepsilon \right\},$$

where $\varepsilon > 0$ and f is any bounded continuous functions on the space X. We will say that a sequence of measures P_n *converges weakly* to the measure P if convergence takes place in the weak topology; in other words, P_n converges weakly to P if for any bounded continuous function f,

$$\int_X f(x) P_n(dx) \underset{n}{\to} \int_X f(x) P(dx).$$

In the case where X is compact metric, the space $\mathcal{M}(X)$ is also compact with respect to the weak topology.

3. Probability Spaces

An arbitrary set Ω, together with a σ-algebra $\mathcal{A}$ of subsets of Ω and a positive measure P defined on $\mathcal{A}$ and normalized so $P(\Omega) = 1$, is a *probability space*. In speaking of a probability space $(\Omega, \mathcal{A}, P)$, the elements $\omega \in \Omega$ are usually called elementary events, the sets $A \in \mathcal{A}$ are events, and the measure $P(A)$ is the *probability* of event A occurring.

The concept of independence is of fundamental importance. Events $A_1, \ldots, A_n$ are called *independent* if

$$P(A_1 \cap \cdots \cap A_n) = P(A_1) \cdots P(A_n); \tag{1.10}$$

σ-algebras $\mathcal{A}_1, \ldots, \mathcal{A}_n \subseteq \mathcal{A}$ are independent if (1.10) holds for any events $A_1 \in \mathcal{A}_1, \ldots, A_n \in \mathcal{A}_n$.

A measurable function $\xi = \xi(\omega)$, $\omega \in \Omega$, on a probability space $(\Omega, \mathcal{A}, P)$ taking values in a measure space $(X, \mathcal{B})$ is called a *random variable*. The probability measure P^ξ defined on the space X by $P^\xi(B) = P\{\xi \in B\}$, $B \in \mathcal{B}$, is the probability distribution of the random variable ξ.

We will say that the σ-algebra $\mathcal{A}$ is generated by a family of variables ξ if it is generated by all possible events of the form $\{\xi \in B\}$, $B \in \mathcal{B}$. Random variables ξ_1, ξ_2 with values in X will be called equivalent if $P\{\xi_1 \neq \xi_2\} = 0$, in other words if ξ_1 and ξ_2 are equal with probability 1. All random variables equivalent to the random variable ξ have the same probability distribution P^ξ.

Let ξ_k, $k = 1, \ldots, n$, be random variables with values in spaces $(X_k, \mathcal{B}_k)$. Their *joint probability distribution* $P^{\xi_1 \cdots \xi_n}$ is defined as the distribution on the semi-ring of sets of the form $B_1 \times \cdots \times B_n$, $B_i \in \mathcal{B}_i$, in the product space $X_1 \times \cdots \times X_n$ given by

$$P^{\xi_1 \cdots \xi_n}(B_1 \times \cdots \times B_n) = P\{\xi_1 \in B_1, \ldots, \xi_n \in B_n\},$$

for all $B_1 \in \mathcal{B}_1, \ldots, B_n \in \mathcal{B}_n$.

We call random variables $\xi_1, \ldots, \xi_n$ *independent* if

$$P^{\xi_1, \ldots, \xi_n}(B_1 \times \cdots \times B_n) = P^{\xi_1}(B_1) \cdots P^{\xi_n}(B_n), \qquad B_1 \in \mathcal{B}_1, \ldots, B_n \in \mathcal{B}_n.$$

Random Functions. Let $(E, \mathscr{B})$ be a measurable space and T an arbitrary set. The family of random variables $\xi(t)$, $t \in T$, with values in $(E, \mathscr{B})$ is a *random function on the set T with phase space* $(E, \mathscr{B})$. The distributions

$$P_S(B_1 \times \cdots \times B_n) = P\{\xi(t_1) \in B_1, \ldots, \xi(t_n) \in B_n\}, \qquad S = (t_1, \ldots, t_n),$$

on the products E^S are called the finite dimensional distributions of the random function $\xi = \xi(t)$, $t \in T$. Recall that each random variable $\xi(t) = \xi(\omega, t)$, $\omega \in \Omega$, is defined on the probability space (Ω, A, P); for each fixed $\omega \in \Omega$, the function $\xi(\omega, \cdot) = \xi(\omega, t)$, $t \in T$, is called a *trajectory*.

Let E be a compact separable metric space. For a given family of consistent probability distributions P_S, $S \subseteq T$, on E one can define a probability space and a family of random variables $\xi(t)$, $t \in T$, with finite dimensional distributions P_S, $S \subseteq T$: for Ω, take the coordinate space $X = E^T$, and for each $t \in T$ define $\xi(t) = \xi(\omega, t)$ as a function of $\omega = x \in X$ by the equation

$$\xi(\omega, t) = x(t), \quad \text{where} \quad x = \{x(t), t \in T\};$$

the corresponding probability measure P is defined on the σ-algebra $\mathscr{A} = \mathscr{A}(T)$ generated by all cylinder sets by means of the given distributions P_S, $S \subseteq T$, by equation (1.7).

A random function $\xi(t)$, $t \in T$, into the phase space $(E, \mathscr{B})$ gives a measurable mapping

$$\omega \to x = \xi(\omega, t), \qquad t \in T \tag{1.11}$$

from the probability space $(\Omega, \mathscr{A}, P)$ to the function space $X = E^T$ with probability distribution P^ξ on the σ-algebra $\mathscr{A}(T)$.

Random functions into the space E are called *equivalent* if for all $t \in T$, $\xi_1(t)$ and $\xi_2(t)$ are equivalent. The finite dimensional distributions of equivalent random functions coincide. In the class of all equivalent random functions one usually distinguishes a suitable representative having particular properties for the trajectory (that is, with trajectories in a particular function space X).

Let T be a topological space; a random function $\xi(t)$, $t \in T$, into a metric phase space E with distance function ρ is called *stochastically continuous* if for any $\varepsilon > 0$,

$$\lim_{s \to t} P\{\rho(\xi(s), \xi(t)) \geq \varepsilon\} = 0.$$

When speaking of random variables or random functions we will, as a rule, mean real (or complex) valued variables ξ. In this case, we let $E\xi$ denote the *mathematical expectation* of the random variable ξ,

$$E\xi = \int_\Omega \xi(\omega)P(d\omega).$$

We frequently consider the spaces $L^p(\mathscr{A}) = L^p(\Omega, \mathscr{A}, P)$, $p = 1, 2$ of all random variables ξ such that $E|\xi|^p < \infty$, with corresponding norm $\|\xi\| =$

$(E|\xi^p|)^{1/p}$; when $p = 2$ this gives the scalar product $(\xi_1, \xi_2) = E\xi_1 \cdot \overline{\xi_2}$. Convergence in the spaces $L^p(\mathscr{A})$ will be called *convergence in mean* ($p = 1$) and *in mean square* ($p = 2$).

In speaking of random variables $\xi \in L^p(\mathscr{A})$ we will not distinguish between equivalent random variables. In accordance with this we will not distinguish between σ-algebras which differ only by events of probability zero.

Let T be a domain in Euclidean space $\mathbb{R}^d$ and $\xi(t)$, $t \in T$, be a random function with finite second moments $E|\xi(t)|^2 < \infty$. In speaking of continuity, differentiability, or integrability we will mean the existence of these properties for $\xi(t)$, $t \in T$, regarded as a function on T with values in the Hilbert space $L^2(\Omega, \mathscr{A}, P)$.

Random Measures. Let T be a measurable space with a ring of measurable sets $\mathfrak{G}$; to each set $\Delta \in \mathfrak{G}$ associate a real or complex random variable $\eta(\Delta)$ with mean zero, $E\eta(\Delta) = 0$, and finite second moment, $E|\eta(\Delta)|^2 < \infty$. This defines a function on $\mathfrak{G}$ with values in $L^2(\Omega, \mathscr{A}, P)$ which we require to be additive: for disjoint $\Delta_1, \Delta_2 \in \mathfrak{G}$, $\eta(\Delta_1 \cup \Delta_2) = \eta(\Delta_1) + \eta(\Delta_2)$.

Suppose, in addition, that $E\eta(\Delta_1)\overline{\eta(\Delta_2)} = 0$ when Δ_1 and Δ_2 are disjoint and that the real-valued additive function $\mu(\Delta) = E|\eta(\Delta)|^2$ is a continuous distribution on the ring $\mathfrak{G}$. A random function $\eta(\Delta)$, $\Delta \in \mathfrak{G}$, having these properties is usually called a *random* (or *stochastic) orthogonal measure*. To characterize these, we will use the symbolic notation

$$E|\eta(dt)|^2 = \mu(dt).$$

For a measurable function $\varphi(t)$, $t \in T$, square-integrable with respect to the measure $\mu(dt)$, a standard construction defines the *stochastic integral*

$$\int_T \varphi(t)\eta(dt) \in L^2(\Omega, \mathscr{A}, P)$$

having the property that

$$E \int_T \varphi(t)\eta(dt) = 0,$$

$$E \left| \int_T \varphi(t)\eta(dt) \right|^2 = \int_T |\varphi(t)|^2 \mu(dt),$$

and

$$E\left[\int_T \varphi_1(t)\eta(dt) \right]\left[\overline{\int_T \varphi_2(t)\eta(dt)} \right] = \int_T \varphi_1(t)\overline{\varphi_2(t)}\mu(dt).$$

Generalized Random Functions. Let T be an open domain in d-dimensional Euclidean space $\mathbb{R}^d$ and $C_0^\infty(T)$ the space of infinitely differentiable functions $u = u(t)$, $t \in T$, with compact support Supp $u \subseteq T$. We can regard $C_0^\infty(T)$ as the union of topological spaces $C_0^\infty(T_{\text{loc}})$, T_{loc} a compact subset of T, each having a neighborhood basis at the origin of the form $\{u \colon \|u\|_t < \varepsilon\}$; here

$$\|u\|_l^2 = \sum_{|k| \le l} \|D^k u\|^2, \qquad l = 0, 1, \ldots,$$

and

$$\|D^k u\|^2 = \int_T |D^k u|^2 \, dt,$$

where

$$D^k u(t) = \frac{\partial^{|k|} u(t)}{\partial t_1^{k_1} \cdots \partial t_d^{k_d}}, \qquad t = (t_1, \ldots, t_d) \in \mathbb{R}^d,$$

$$k = (k_1, \ldots, k_d) \quad \text{and} \quad |k| = k_1 + \cdots + k_d.$$

Convergence of a sequence $u_n \to u$ in the space $C_0^\infty(T)$ means that the functions u_n all have support Supp $u_n \subseteq T_{\text{loc}}$ for some compact $T_{\text{loc}} \subseteq T$ and that $u_n \to u$ in the topological space $C_0^\infty(T_{\text{loc}})$.

Consider a continuous linear map from the space $C_0^\infty(T)$ into $L^2(\Omega, \mathscr{A}, P)$, under which the functions $u \in C_0^\infty(T)$ correspond to random variables denoted by $(u, \xi) \in L^2(\Omega, \mathscr{A}, P)$. We will call this continuous linear operator $\xi = (u, \xi)$, $u \in C_0^\infty(T)$, a *generalized random function*. For $\xi = (u, \xi)$, we define the operations differentiation, multiplication by a C^∞ function, etc., as they are usually defined for ordinary generalized functions; that is,

$$D^k \xi = (-1)^{|k|}(D^k u, \xi),$$

$a \cdot \xi = (\bar{a} \cdot u, \xi)$ for $a = a(t)$, $t \in T$, an infinitely differentiable function.

An example of a generalized random function is given by the operator $(u, \xi) = \int_T u(t)\overline{\xi(t)} \, dt$, $u \in C_0^\infty(T)$, where the function $\xi(t) \in L^2(\Omega, \mathscr{A}, P)$, $t \in T$, is required to be integrable on every bounded domain $S \subseteq T$ and in particular,

$$\|(u, \xi)\| \le \int_T |u(t)| \, \|\xi(t)\| \, dt.$$

It is in the above sense that we will speak of a generalized random function henceforth.

Another example is offered by so-called "white noise" $\dot{\eta}(t)$, $t \in T$; this is a generalized random function of the form

$$(u, \dot{\eta}) = \int_T u(t)\dot{\eta}(t) \, dt = \int_T u(t)\eta(dt), \qquad u \in C_0^\infty(T).$$

The first expression only makes sense when interpreted according to the second (stochastic) integral, in which $\eta(dt)$ is the orthogonal random measure for which $E|\eta(dt)|^2$ is the Lebesgue measure dt.

§2. Conditional Distributions

1. Conditional Expectation

We turn to the Hilbert space $L^2(\mathcal{A}) = L^2(\Omega, \mathcal{A}, P)$, whose elements are random variables ξ, $E|\xi|^2 < \infty$, with scalar product $(\xi_1, \xi_2) = E\xi_1\bar{\xi}_2$; we stipulate that equivalent random variables be represented by a single element in this space.

Let $\mathcal{B}$ be some σ-algebra of events, $\mathcal{B} \subseteq \mathcal{A}$ and $L^2(\mathcal{B})$ be the subspace of all random variables $\eta \in L^2(\mathcal{A})$ which are measurable relative to $\mathcal{B}$. For each variable $\xi \in L^2(\mathcal{A})$ we let

$$E(\xi|\mathcal{B}) = P(\mathcal{B})\xi, \tag{2.1}$$

where $P(\mathcal{B})$ is the orthogonal projection from $L^2(\mathcal{A})$ onto $L^2(\mathcal{B})$. The projection $E(\xi|\mathcal{B})$ is the closest approximation to ξ among all variables $\eta \in L^2(\mathcal{B})$:

$$\|\xi - E(\xi|\mathcal{B})\| = \min_{\eta \in L^2(\mathcal{B})} \|\xi - \eta\|,$$

and in view of this one can interpret the random variable $E(\xi|\mathcal{B})$ as the best estimate for ξ when all known outcomes are events $B \in \mathcal{B}$.

We introduce the indicator functions

$$1_A(\omega) = \begin{cases} 1, & \omega \in A, \\ 0, & \omega \notin A. \end{cases}$$

The random variables 1_B, $B \in \mathcal{B}$, form a complete basis of the subspace $L^2(\mathcal{B})$ and the projection $E(\xi|\mathcal{B}) \in L^2(\mathcal{B})$ is uniquely defined by the orthogonality conditions $(\xi - E(\xi|\mathcal{B}), 1_B) = 0$, $B \in \mathcal{B}$, which we can write in the form

$$\int_B \xi(\omega)P(d\omega) = \int_B E(\xi|\mathcal{B})P(d\omega), \qquad B \in \mathcal{B}. \tag{2.2}$$

When $B = \Omega$ this gives the equation

$$E(E(\xi|\mathcal{B})) = E\xi. \tag{2.3}$$

From condition (2.2) it is immediately obvious that

$$E(\xi|\mathcal{B}) \geq 0 \quad \text{a.e. for } \xi \geq 0, \tag{2.4}$$

since for $B = \{E(\xi|\mathcal{B}) < 0\}$, equation (2.2) is satisfied only if $P(B) = 0$.

In the following discussion we consider random variables as elements of the space $L^2(\mathcal{A})$ without distinguishing between variables which agree almost everywhere.

From (2.4) we have

$$|E(\xi|\mathcal{B})| \leq E(|\xi| \,|\, \mathcal{B}). \tag{2.5}$$

In the same way, using the fact that $E(1|\mathscr{B}) = 1$, we have for any random variable ξ

$$c_1 \leq E(\xi|\mathscr{B}) \leq c_2 \quad \text{when } c_1 \leq \xi \leq c_2. \tag{2.6}$$

Let $\eta \in L^2(\mathscr{B})$ and suppose one of the variables ξ, η is bounded so that $\xi \cdot \eta \in L^2(\mathscr{A})$; then

$$E(\xi \cdot \eta|\mathscr{B}) = \eta \cdot E(\xi|\mathscr{B}) \tag{2.7}$$

because of the orthogonality conditions:

$$(\eta \cdot \xi - \eta \cdot E(\xi|\mathscr{B}), 1_B) = (\xi - E(\xi|\mathscr{B}), \eta \cdot 1_B)$$
$$= 0, \quad B \in \mathscr{B}.$$

Let $\mathscr{F} \subseteq \mathscr{A}$ be a σ-algebra containing $\mathscr{B}$; then

$$E(\xi|\mathscr{B}) = E(E(\xi|\mathscr{F})|\mathscr{B}), \tag{2.8}$$

since for the corresponding projection operators we have $P(\mathscr{B}) = P(\mathscr{B})P(\mathscr{F})$.

Now we consider a family of σ-algebras $\mathscr{B}(S)$, depending on a partially ordered set of parameters S. With later applications in mind, we assume $S \subseteq T$, with T a separable locally-compact metric space.

Let $\mathscr{B}(S_1) \supseteq \mathscr{B}(S_2)$ when $S_1 \supseteq S_2$.

We will look at the family $\mathscr{B}(S)$ directed by inclusion $\supseteq$, where the parameters S are the complements of compacts in the space T. For such a directed family there exists the limit

$$\lim E(\xi|\mathscr{B}(S)) = E(\xi|\mathscr{B}), \quad \mathscr{B} = \bigcap \mathscr{B}(S). \tag{2.9}$$

In fact, there is a sequence $S_1 \supseteq S_2 \supseteq \cdots$, such that for any set S which is the complement of a compact set, we can find $S_n \subseteq S$; we call such a sequence *directing* for our directed family. We have $\mathscr{B} = \bigcap_n \mathscr{B}(S_n)$ and $\bigcap L^2(\mathscr{B}(S)) = \bigcap_n L^2(\mathscr{B}(S_n)) = L^2(\mathscr{B})$. Let $\eta_S = E(\xi|\mathscr{B}(S))$ be the projections of the random variable ξ on the subspaces $L^2(\mathscr{B}(S))$. Clearly the limit $\eta = \lim_{n \to \infty} \eta_{S_n}$ exists, where $\eta = E(\xi|\mathscr{B})$ is the projection on $L^2(\mathscr{B})$; for any $S \subseteq S_n$ we have $L^2(\mathscr{B}(S)) \subseteq L^2(\mathscr{B}(S_n))$ and thus $\|\eta_S - \eta\| \leq \|\eta_{S_n} - \eta\|$.

An analogous result is also true for the family $\mathscr{B}(S)$ directed by inclusion $\subseteq$. Namely, take a directing sequence $S_1 \subseteq S_2 \subseteq \cdots$ such that for any S there is some $S_n \supseteq S$. Then the σ-algebra $\mathscr{B}$ generated by the given family $\mathscr{B}(S)$ is also generated by the sequence of increasing σ-algebras $\mathscr{B}(S_n)$: $\mathscr{B} = \bigvee \mathscr{B}(S) = \bigvee_n \mathscr{B}(S_n)$. The sequence of random variables $\eta_n = \eta_{S_n}$, satisfies $\|\eta_n\|^2 \leq \|\xi\|^2$ and

$$\|\eta_n - \eta_m\|^2 \leq \|\eta_n\|^2 - \|\eta_m\|^2 \to 0, \quad n \geq m \to \infty,$$

and has a limit random variable η for which the difference $\xi - \eta$ is orthogonal to all $L^2(\mathscr{B}(S_n))$, since $\eta = E(\xi|\mathscr{B})$. It follows that the limit

$$\lim E(\xi|\mathscr{B}(S)) = E(\xi|\mathscr{B}), \quad \mathscr{B} = \bigvee \mathscr{B}(S) \tag{2.10}$$

exists.

We should emphasize that in both limits (2.9) and (2.10), it is a question of convergence in mean square; however, if we confine ourselves to a directing sequence S_n, $n = 1, 2, \ldots$, then it is known that we also have convergence almost everywhere—a nontrivial result, in contrast to the question of convergence in mean square.

Now we consider the space $L^1(\mathscr{A}) = L^1(\Omega, \mathscr{A}, P)$, and regard $L^2(\mathscr{A})$ as a subspace of $L^1(\mathscr{A})$ whose completion with respect to the norm $\|\xi\| = E|\xi|$ coincides with $L^1(\mathscr{A})$. From (2.3) and (2.5) it is apparent that the linear operator $E(\xi|\mathscr{B})$ defined on random variables $\xi \in L^2(\mathscr{A})$ is bounded in $L^1(\mathscr{A})$ norm,

$$\|E(\xi|\mathscr{B})\| \leq \|\xi\|,$$

and by continuity it can be uniquely extended to the whole space $L^1(\mathscr{A})$. It is also clear that by the continuity of the operator $E(\cdot/\mathscr{B})$ thus defined, properties established earlier for it also hold for variables $\xi \in L^1(\mathscr{A})$.

The linear operator $E(\cdot|\mathscr{B})$ is called *conditional expectation*; more precisely $E(\xi|\mathscr{B})$ is the conditional expectation of the random variable $\xi \in L^1(\mathscr{A})$.

For variables $\xi \in L^1(\mathscr{A})$, the limits (2.9), (2.10) are true for convergence in mean, since by approximating ξ in $L^1(\mathscr{A})$ by variables $\eta \in L^2(\mathscr{A})$ we have

$$\|E(\xi|\mathscr{B}(S)) - E(\xi|\mathscr{B})\| \leq \|E(\eta|\mathscr{B}(S)) - E(\eta|\mathscr{B})\| + 2\|\xi - \eta\|.$$

2. Conditional Probability Distributions

Let 1_A be the indicator function of an event $A \in \mathscr{A}$. The conditional expectation

$$P(A|\mathscr{B}) = E(1_A|\mathscr{B}) \qquad (2.11)$$

is the *conditional probability* of A with respect to the σ-algebra $\mathscr{B}$; it can be interpreted as the probability of the event A when the outcomes of all events $B \in \mathscr{B}$ are known. In particular

$$P(B|\mathscr{B}) = 1_B, \qquad B \in \mathscr{B}.$$

Clearly,

$$P(\Omega|\mathscr{B}) = 1,$$
$$P(A|\mathscr{B}) = 0, \quad \text{if } P(A) = 0, \qquad (2.12)$$

and by (2.6), for any $A \in \mathscr{A}$, $0 \leq P(A|\mathscr{B}) \leq 1$. In fact for any disjoint events $A_1, A_2, \ldots \in \mathscr{A}$ and $A = \bigcup_k A_k$ we have

$$P(A|\mathscr{B}) = \sum_k P(A_k|\mathscr{B}) \qquad (2.13)$$

from the continuity of the operator $E(\cdot|\mathscr{B})$, since

$$1_A = \sum_k 1_{A_k} = \lim_{n\to\infty} \sum_{k\le n} 1_{A_k}$$

in the space $L^1(\mathscr{B})$.

With the help of conditional probability, one can define a different type of relationship between events. Events $A_1, \ldots, A_n$ are called *conditionally independent with respect to the σ-algebra $\mathscr{B}$* if

$$P(A_1 \cdot\ldots\cdot A_n|\mathscr{B}) = P(A_1|\mathscr{B}) \cdots P(A_n|\mathscr{B}), \tag{2.14}$$

(cf. (1.10)). In the same way we define conditional independence of σ-algebras $\mathscr{A}_1, \ldots, \mathscr{A}_n$, requiring that (2.14) hold for any events $A_1 \in \mathscr{A}_1, \ldots, A_n \in \mathscr{A}_n$.

A sequence $\mathscr{A}_1, \mathscr{A}_2, \ldots$ is called *Markov* if relative to a "past", $\bigvee_{i\le j} \mathscr{A}_i$, we have for any $k > j$,

$$P\left(A_k \,\bigg|\, \bigvee_{i\le j} \mathscr{A}_i\right) = P(A_k|\mathscr{A}_j), \qquad A_k \in \mathscr{A}_k;$$

we remark that this equality can be extended from events A_k to all events in the "future":

$$P\left(A \,\bigg|\, \bigvee_{i\le j} \mathscr{A}_i\right) = P(A|\mathscr{A}_j), \qquad A \in \bigvee_{k>j} \mathscr{A}_k.$$

We can regard conditional probability as a function of $\omega \in \Omega$, denoting this by

$$P(A, \omega) = P(A|\mathscr{B});$$

this function is measurable with respect to the σ-algebra $\mathscr{B}$ and among all such functions in $L^1(\mathscr{A})$ is uniquely defined for almost every ω by the equality

$$\int_B P(A, \omega)P(d\omega) = P(AB), \qquad B \in \mathscr{B},$$

(cf. (2.2)). If for almost every fixed $\omega \in \Omega$, the value of $P(A, \omega)$ is chosen so that the set function

$$P(A, \omega) = P(A|\mathscr{B}), \qquad A \in \mathscr{A}, \tag{2.15}$$

is a probability measure, then the measures (2.15) are called the *conditional probability distributions* with respect to the σ-algebra $\mathscr{B}$.

Existence of Conditional Distributions. The question here is whether it is possible to define for each $A \in \mathscr{A}$ a version of the conditional probability $P(A|\mathscr{B})$ among all the candidates, $\mathscr{B}$-measurable and agreeing almost everywhere, so that for the various A, condition (2.13) will hold for each given ω, for almost every $\omega \in \Omega$.

Conditional distributions exist on the σ-algebra $\mathscr{A}$, if $\mathscr{A}$ is separable and the probability measure P is perfect. We will show this.

The σ-algebra $\mathscr{A}$ is generated by a countable collection of sets $\mathfrak{G}$ and we can immediately assume $\mathfrak{G}$ is an algebra. We index the sets in $\mathfrak{G}$, $A_1, A_2, \ldots$ and consider the real random variable

$$\xi = \sum_n \frac{2}{3^n} 1_{A_n}.$$

Notice that $A_n = \{\xi \in B_n\}$, where B_n is the Borel set on the real line of the form $B_n = \bigcup_k [2k/3^n, (2k + 1)/3^n]$. Let

$$F^\xi(x, \omega) = P\{\xi \leq x | \mathscr{B}\}$$

for all rational $x \in \mathbb{R}$. Consistent with property (2.13), the relations

$$\lim_{x \to -\infty} F^\xi(x, \omega) = 0, \qquad \lim_{x \to \infty} F^\xi(x, \omega) = 1,$$

$$F^\xi(x, \omega) \leq F^\xi(y, \omega), \qquad x < y,$$

$$\lim_{y \to x + 0} F^\xi(y, \omega) = F^\xi(x, \omega),$$

can fail to hold on at most a set of measure 0; that is, there exists a set N with $P(N) = 0$ such that for every $\omega \notin N$, the relations above hold for all rational x and y. As a function of x, for fixed $\omega \notin N$, $F^\xi(x, \omega)$ can be extended by continuity to the whole real line. Furthermore, letting $y \to x + 0$ through the rationals,

$$F^\xi(x, \omega) = \lim_{y \to x + 0} F^\xi(y, \omega) = \lim_{y \to x + 0} P\{\xi \leq y | \mathscr{B}\} = P\{\xi \leq x | \mathscr{B}\}$$

for almost every ω; thus the distribution functions $F^\xi(\cdot, \omega)$ give conditional probability distributions on $\mathbb{R}$ for the random variable ξ, $P^\xi(\cdot, \omega)$, for each $\omega \notin N$.

Now we use the fact that the probability measure P is perfect. Under the mapping $\xi = \xi(\omega)$ the set $\xi(\Omega)$ is measurable and there is a Borel set on the real line, $B_0 \subseteq \xi(\Omega)$, such that

$$P^\xi(B_0) = \int_\Omega P^\xi(B_0, \omega)P(d\omega) = 1.$$

Hence $P^\xi(B_0, \omega) = 1$ for almost all ω, the set $\xi(\Omega)$ is measurable with respect to the conditional probability distributions, and $P^\xi(\xi(\Omega), \omega) = 1$. We know (see (1.9)) that for all such ω we can define a probability measure on the σ-algebra $\mathscr{A} = \mathscr{A}_\xi$ of events of the form $A = \{\xi \in B\}$, B a Borel set in $\mathbb{R}$, by $P(A, \omega) = P^\xi(B, \omega)$, $A \in \mathscr{A}$; these will be the desired conditional distributions.

The argument above remains valid when extended to the conditional distributions

$$P^\xi(A, \omega) = P\{\xi \in A | \mathscr{B}\}, \qquad A \in \mathscr{A}$$

for a random variable ξ with values in the space $(X, \mathscr{A})$. In particular, the separability condition on the σ-algebra $\mathscr{A}$ is satisfied and the probability measure $P = P^\xi$ is perfect if X is a complete separable metric space with Borel σ-algebra $\mathscr{A}$.

To complete this discussion, suppose the σ-algebra $\mathscr{B}$ is generated by a random variable η with values in a measurable space $(Y, \mathscr{B})$; (by abuse of notation, we designate the image σ-algebra by $\mathscr{B}$ also). Then we can choose the conditional probabilities (2.15) to have the form

$$P(A, \omega) = P(A | \eta(\omega)), \qquad A \in \mathscr{A}, \tag{2.16}$$

where the right side is a $\mathscr{B}$-measurable function $P(A|y)$ of $y \in Y$, defined on the measurable set $\eta(\Omega) \subseteq Y$.

EXAMPLE (Conditional Densities). Let ξ_1, ξ_2 be random variables into the measure spaces $(X_1, \mathscr{A}_1, Q_1)$ and $(X_2, \mathscr{A}_2, Q_2)$ such that their joint probability distribution P, on the σ-algebra $\mathscr{A}$ generated by the semi-ring $\mathscr{A}_1 \times \mathscr{A}_2$ in the product $X = X_1 \times X_2$, is absolutely continuous with respect to the measure $Q = Q_1 \times Q_2$ and has density $p(x) = P(dx)/Q(dx), x = (x_1, x_2) \in X$. Then the conditional distribution (2.16) of the variable $\xi = \xi_1$ with respect to $\eta = \xi_2$ can be given by the formula

$$P(\xi_1 \in A | \xi_2) = \int_A p_1(x_1, \xi_2) Q_1(dx_1), \qquad A \in \mathscr{A}_1,$$

$$p_1(x) = \frac{p(x)}{\int_{X_1} p(x) Q_1(dx_1)}. \tag{2.17}$$

EXAMPLE (Markov Sequence). In the preceding example, suppose the variable ξ_2 consists of two components which we indicate by (ξ_2, ξ_3), and replace x_2 by (x_2, x_3) where appropriate. Suppose the conditional density of the variable ξ_1, conditioned on $\xi_2 = x_2, \xi_3 = x_3$, (as defined in (2.17)) does not depend on x_3:

$$p_1(x) = p_1(x_1, x_2), \qquad x = (x_1, x_2, x_3).$$

This is equivalent to the following multiplicative representation of the unconditional density $p(x) = P(dx)/Q(dx)$:

$$p(x) = p_1(x_1, x_2) p_2(x_2, x_3), \tag{2.18}$$

where

$$p_2(x_2, x_3) = \int_{X_1} p(x) Q_1(dx_1).$$

Whenever the density $p(x)$ is of the form (2.18) the sequence (ξ_1, ξ_2, ξ_3) is *Markov*, since

$$P(\xi_1 \in A | \xi_2, \xi_3) = P(\xi_1 \in A | \xi_2), \qquad A \in \mathscr{A}_1.$$

Choice of Conditional Distributions. For any σ-algebra $\mathscr{B}' = \mathscr{B}$ (mod 0) one can choose conditional distributions (2.15) so that they are measurable with respect to $\mathscr{B}'$ and also defined on a set $\Omega' \in \mathscr{B}'$ of full measure, $P(\Omega') = 1$.

When the σ-algebra $\mathscr{A}$ is separable, one can choose such a $\mathscr{B}' = \mathscr{B}$ (mod 0) by looking at the σ-algebra generated by a countable collection of random variables which are complete in the space $L^2(\mathscr{B})$, a subspace of the separable space $L^2(\Omega, \mathscr{A}, P)$. By taking the conditional distributions (2.15) measurable with respect to the separable σ-algebra $\mathscr{B}'$, we get that

$$P(B, \omega) = \begin{cases} 1, & \omega \in B, \\ 0, & \omega \notin B, \end{cases} \quad B \in \mathscr{B}', \quad (2.19)$$

for all $\omega \in \Omega'$, a $\mathscr{B}'$-measurable set with $P(\Omega') = 1$.

In fact, the σ-algebra $\mathscr{B}'$ is generated by a countable algebra of sets B, and for each of these, the function $P(B, \omega)$, regarded as conditional probability, satisfies the equality

$$P(B, \omega) = 1_B(\omega)$$

for all $\omega \in \Omega'$. For fixed ω, both sides of the equation are measures on the σ-algebra $\mathscr{B}$. This can be extended from the algebra of sets generating $\mathscr{B}'$ to every $B \in \mathscr{B}'$.

It is clear from (2.19) that for elements $\omega_1 \neq \omega_2$ in Ω' we have orthogonal measures

$$P(\cdot, \omega_1) \perp P(\cdot, \omega_2), \quad (2.20)$$

provided ω_1, ω_2 are separated by the σ-algebra $\mathscr{B}'$, i.e., one can find disjoint sets $B_1, B_2 \in \mathscr{B}'$ such that $\omega_i \in B_i$; otherwise

$$P(A, \omega_1) = P(A, \omega_2), \quad A \in \mathscr{A}, \quad (2.21)$$

since for every A the function $P(A, \omega)$ is measurable with respect to the σ-algebra $\mathscr{B}'$.

Consistency of Conditional Distributions. Let $\mathscr{E}$ be an arbitrary separable σ-algebra containing $\mathscr{B}$ and let $P(A|\mathscr{E})$ be a conditional probability defined for the probability measure P. It turns out that $P(A|\mathscr{E})$ is a conditional probability for almost all conditional distributions (2.15); this might seem paradoxical in view of the orthogonality property in (2.20).

We will show that

$$P(AC, \omega') = \int_C P(A|\mathscr{E})P(d\omega, \omega'), \quad A \in \mathscr{A}, C \in \mathscr{E} \quad (2.22)$$

for almost all ω'.

The conditional expectation $E(\xi|\mathscr{B})$ can be expressed as a function of $\omega' \in \Omega$ by the integral

$$E(\xi|\mathscr{B}) = \int_\Omega \xi(\omega)P(d\omega, \omega');$$

using the general formula (2.8) we have the following equality:

$$P(A, \omega') = E(1_A | \mathscr{B}) = E(E(1_A | \mathscr{E})\mathscr{B}) = \int_\Omega P(A | \mathscr{E})P(d\omega, \omega')$$

for almost all ω'. Replacing A by the product AC, $C \in \mathscr{E}$, and observing that $P(AC | \mathscr{E}) = 1_C \cdot P(A | \mathscr{E})$, we get (2.22) for fixed A and C. This can be extended to a countable collection of events $A \in \mathfrak{G}(\mathscr{A})$ and $C \in \mathfrak{G}(\mathscr{E})$. Let $\Omega' \in \mathscr{B}$ be that set of measure 1 on which (2.22) is true for all $A \in \mathfrak{G}(\mathscr{A})$, $C \in \mathfrak{G}(\mathscr{E})$. The left-hand side of (2.22), as a function of $A \in \mathscr{A}$, is a measure, and if $P(\cdot | \mathscr{E})$ in (2.22) is taken to be a conditional distribution, then the right-hand side of (2.22) is also a measure. Suppose $\mathfrak{G}(\mathscr{A})$ is an algebra which generates the σ-algebra $\mathscr{A}$; then for every fixed $\omega' \in \Omega'$ equation (2.22) extends to all events $A \in \mathscr{A}$. Equation (2.22) can be extended from the algebra $\mathfrak{G}(\mathscr{E})$ to all events $C \in \mathscr{E}$ for the same reasons.

§3. Zero-One Laws. Regularity

1. Zero-one Law

We say that a sequence of σ-algebras $\mathscr{A}_1 \supseteq \mathscr{A}_2 \supseteq \cdots$ satisfies a *zero-one law* if the intersection $\mathscr{A}_\infty = \bigcap_n \mathscr{A}_n$ is trivial, consisting only of events of probability 0 or 1.

A classic example of the operation of a zero-one law is in connection with a sequence of independent random variables $\xi_1, \xi_2, \ldots$, where the σ-algebras $\mathscr{A}_n$ are generated by the corresponding "tails" $\xi_n, \xi_{n+1}, \ldots$ of the sequence.

Later we will consider a more general scheme, in which we take a family of random variables $\xi(t)$, $t \in T$, depending on a parameter $t \in T$, together with sequences $\mathscr{A}_1 \supseteq \mathscr{A}_2 \supseteq \cdots$ arising in a particular way from a family $\mathscr{A}(S)$, $S \subseteq T$, of σ-algebras of events generated by the variables $\xi(t)$, $t \in S$, where the σ-algebras are directed by inclusion $\supseteq$.

Regularity. We look at a collection $\mathscr{A}(S)$, $S \subseteq T$, where each σ-algebra of events $\mathscr{A}(S)$ is associated with a set S of parameters in the space T. We will assume that T is a separable, locally-compact metric space and that the σ-algebras $\mathscr{A}(S)$, $S \subseteq T$, satisfy the following conditions:

$$\mathscr{A}(S_1) \supseteq \mathscr{A}(S_2) \quad \text{when } S_1 \supseteq S_2 \tag{3.1}$$

and secondly

$$\mathscr{A}(S) = \bigvee_{S' \subseteq S} \mathscr{A}(S') \tag{3.2}$$

is generated by events from all possible compact domains $S' \subseteq S$.†

† By a *domain* we mean a set contained in the closure of its interior points.

We remark here that condition (3.2) will be used only in connection with the concluding examples in §1.

We let

$$\mathscr{A}(\infty) = \bigcap \mathscr{A}(S), \tag{3.3}$$

where the right side is taken over a family of domains $S \subseteq T$ which are complements of compacts in the space T and which are directed by inclusion $\supseteq$. We call the collection $\mathscr{A}(S)$, $S \subseteq T$, *regular* if the σ-algebra $\mathscr{A}(\infty)$ is trivial.

Now consider the Hilbert space $L^2(\Omega, \mathscr{A}, P)$, $\mathscr{A} = \mathscr{A}(T)$, and subspaces $L^2(\mathscr{A}(S))$, each consisting of random variables measurable with respect to the corresponding σ-algebra $\mathscr{A}(S) \subseteq \mathscr{A}$. It is clear that

$$L^2(\mathscr{A}(\infty)) = \bigcap_n L^2(\mathscr{A}(S_n)),$$

where $S_n, n = 1, 2, \ldots$, is a directed sequence for the given family of domains S consisting of complements of compacts. The regularity condition implies that the space $L^2(\mathscr{A}(\infty))$ consists only of constant functions and is equivalent to saying that for some $L^2(\mathscr{A}(T))$-complete set of random variables ξ,

$$\lim_{n \to \infty} E(\xi \,|\, \mathscr{A}(S_n)) = E\xi. \tag{3.4}$$

(Recall that $E(\xi \,|\, \mathscr{A}(S_n)) = \xi_n$ is the projection of the variable ξ on $L^2(\mathscr{A}(S_n))$.) Take $\xi = 1_A - P(A)$, $A \in \mathscr{A}$; condition (3.4) means that $\lim_{n \to \infty} \|\xi_n\| = 0$. For any variable $\eta \in L^2(\mathscr{A}(S_n))$ we have

$$|(\xi, \eta)| = |(\xi_n, \eta)| \le \|\xi_n\| \cdot \|\eta\|,$$

and for the set of variables $\eta = 1_B - P(B)$, $B \in \mathscr{A}(S_n)$, bounded in norm, we have $(\xi, \eta) = P(AB) - P(A)P(B)$ and the regularity condition implies the limit relation:

$$\lim_{\substack{n \to \infty \\ B \in \mathscr{A}(S_n)}} \sup |P(AB) - P(A)P(B)| = 0 \quad \text{for any } A \in \mathscr{A}. \tag{3.5}$$

Condition (3.5) is not only necessary but sufficient for regularity since for $A = B \in \mathscr{A}(\infty)$ it gives the equation $P(A) = P(A)^2$ which can hold only when $P(A) = 0$ or 1.

Notice that in (3.5) we can limit ourselves to considering events A in some algebra $\mathfrak{G}$ generating the σ-algebra $\mathscr{A}(T)$, since each event $A \in \mathscr{A}(T)$ can be approximated by events $A_\varepsilon \in \mathfrak{G}$: for each $\varepsilon > 0$ there is an $A_\varepsilon \in \mathfrak{G}$ such that $\|1_A - 1_{A_\varepsilon}\| = P(A \circ A_\varepsilon) \le \varepsilon$; in the same way we can restrict ourselves to events B in the algebra generating the σ-algebra $\mathscr{A}(S_n)$. In particular, one can consider only events $A \in \mathscr{A}(S')$ and $B \in \mathscr{A}(S'')$ corresponding to all appropriate compact sets $S' \subseteq T$, $S'' \subseteq S_n$. This allows us to express the regularity condition in the following form: for each event $A \in \mathscr{A}(S')$, S' compact in T,

$$\lim_{\substack{n \to \infty \\ A'' \in \mathscr{A}(S'')}} \sup |P(A' \cdot A'') - P(A')P(A'')| = 0, \tag{3.6}$$

where the sup is taken over all A'' in the σ-algebra corresponding to compact $S'' \subseteq S_n$.

Clearly condition (3.6) is satisfied if the σ-algebras $\mathscr{A}(S')$ and $\mathscr{A}(S'')$ correspond to disjoint compact sets and are independent. In general (3.6) can be interpreted as weak dependence of events in σ-algebras $\mathscr{A}(S')$ and $\mathscr{A}(S'')$ corresponding to S' and S'' which are "far apart."

2. Decomposition Into Regular Components

We turn our attention to the collection $\mathscr{A}(S)$, $S \subseteq T$, considered in §3.1, assuming that all σ-algebras $\mathscr{A}(S)$ are separable and the probability measure P on the space $(\Omega, \mathscr{A})$, $\mathscr{A} = \mathscr{A}(T)$, is perfect.

We complete the collection $\mathscr{A}(S)$, $S \subseteq T$, by adding $\mathscr{A}(\infty)$, defined in (3.3). Applying our previous results to the probability space $(\Omega, \mathscr{A}, P)$, there exist conditional probability distributions

$$P_S(A, \omega) = P(A \mid \mathscr{A}(S)), \qquad A \in \mathscr{A}, \tag{3.7}$$

including one for $S = \{\infty\}$.

Let S_n, $n = 1, 2, \ldots$, be a directed sequence in the family of domains $S \subseteq T$ which are complements of compacts in T; the domains are directed by inclusion $\supseteq$. We know that for each fixed $A \in \mathscr{A}$ the limit exists for almost all ω:

$$\lim P_{S_n}(A, \omega) = P_\infty(A, \omega). \tag{3.8}$$

We choose conditional distributions $P_\infty(A, \omega) = P(A \mid \mathscr{A}(\infty))$, $A \in \mathscr{A}$, measurable with respect to a separable σ-algebra $\mathscr{B} = \mathscr{A}(\infty)$ (mod 0), defining them on a set $\Omega_\infty \in \mathscr{B}$ of P-measure 1 so that first, the probability measures

$$Q = P_\infty(\cdot, \omega), \qquad \omega \in \Omega_\infty, \tag{3.9}$$

will have the properties listed in (2.19)–(2.21); second, the limit relationship (3.8) will hold for every $\omega \in \Omega_\infty$ for each $A \in \mathfrak{G}(\mathscr{A})$, $\mathfrak{G}(\mathscr{A})$ a countable algebra generating the σ-algebra $\mathscr{A}$; and third, for all $S = S_n, n = 1, 2, \ldots$, the distributions (3.7) are conditional distributions with respect to each probability measure Q (c.f. (2.22)).

In accordance with (2.19), for each event $B \in \mathscr{B}$ we have

$$P_\infty(B, \omega') = \begin{cases} 1, & \omega' \in B, \\ 0, & \omega' \notin B. \end{cases}$$

We will show that the collection $\mathscr{A}(S)$, $S \subseteq T$, is regular with respect to every probability measure $Q = P_\infty(\cdot, \omega)$.

Since $\omega' \in \Omega_\infty$ and $\Omega_\infty \in \mathscr{B}$, we have $Q(\Omega_\infty) = 1$, and the limit relation (3.8) gives us the regularity condition (3.4) for the probability measure Q. In fact, if we take a complete system of variables $\xi = 1_A$, $A \in \mathfrak{G}(\mathscr{A})$, in the space $L^2(\Omega, \mathscr{A}, Q)$ we see that the corresponding projections

$$E(\xi \mid \mathscr{A}(S_n)) = P_{S_n}(A, \omega)$$

converge to a function $\phi(\omega) = P_\infty(A, \omega)$ on the set Ω_∞ of full Q-measure, $Q(\Omega_\infty) = 1$. The function $\varphi(\omega)$ is measurable with respect to the σ-algebra $\mathscr{B}$ and consequently $\varphi(w)$ is constant almost everywhere with respect to Q:

$$\varphi(\omega) = E\xi = Q(A).$$

From this it follows that the regularity condition for the directed collection $\mathscr{A}(S)$, $S \subseteq T$, is satisfied with respect to all probability measures (3.9) and the expression

$$P(A) = \int P(A, \omega)P(d\omega), \qquad A \in \mathscr{A}, \tag{3.10}$$

can be regarded as a decomposition of the probability measure P into "regular components" $P(\cdot, \omega) = P_\infty(\cdot, \omega)$.

§4. Consistent Conditional Distributions

1. Consistent Conditional Distributions for a Given Probability Measure

We consider again the collection of σ-algebras $\mathscr{A}(S)$ depending on sets $S \subseteq T$ as parameters and satisfying (3.1): $\mathscr{A}(S_1) \supseteq \mathscr{A}(S_2)$ for $S_1 \supseteq S_2$.

Let $\mathscr{S}$ be a system of sets $S \subseteq T$ which are partially ordered by inclusion $\supseteq$, and let

$$P_S(A, \omega) = P(A \mid \mathscr{A}(S)), \qquad A \in \mathscr{A}, \tag{4.1}$$

be a family of conditional distributions on the probability space $(\Omega, \mathscr{A}, P)$ depending on $\omega \in \Omega_S$ and $S \in \mathscr{S}$, where $\Omega_S \in \mathscr{A}(S)$ with $P(\Omega_S) = 1$.

For each $\omega' \in \Omega_{S_2}$, let Q be the probability measure $P_{S_2}(\cdot, \omega')$. The conditional distributions (4.1) will be called *consistent* if for $S_1 \supseteq S_2$ and for each $\omega' \in \Omega_{S_2}$,

$$P_{S_1}(A, \omega) = Q(A \mid \mathscr{A}(S_1)), \qquad A \in \mathscr{A}. \tag{4.2}$$

We write the conditional probability on the right-hand side as $P_{S_2}(A \mid \mathscr{A}(S_1))$. Note that the consistency condition implies the equality

$$P_{S_2}(\Omega_{S_1}, \omega') = 1, \qquad \omega' \in \Omega_{S_2},$$

since

$$P_{S_2}(\Omega_{S_1}, \omega') = Q(\Omega_{S_1}) = Q(\Omega_{S_1} \mid \mathscr{A}(S_1))$$
$$= P_{S_1}(\Omega_{S_1}, \omega) = 1.$$

Assume the probability measure P is perfect and each σ-algebra $\mathscr{A}(S)$, $S \in \mathscr{S}$, is separable and for each S there are only a countable number of distinct $\mathscr{A}(S_\alpha)$, $S_\alpha \supseteq S$. Then one can define consistent conditional distributions (4.1) such that for each $S \in \mathscr{S}$, the distributions $P_S(\cdot, \omega)$, $\omega \in \Omega_S$, have properties (2.19)–(2.21) and such that for any previously specified family S_α ordered by inclusion $\supseteq$, with $\bigcap S_\alpha = S$, the following limit exists

$$\lim P_{S_\alpha}(A, \omega) = P_S(A, \omega), \tag{4.3}$$

for all $\omega \in \Omega_S$ and $A \in \mathfrak{G}(\mathscr{A})$, where $\mathfrak{G}(\mathscr{A})$ is some specified countable algebra generating the σ-algebra $\mathscr{A}$. This is completely analogous to our previous development of (3.9) for conditional probabilities.

2. Probability Measures with Given Conditional Distributions

Suppose we are given probability distributions

$$P_S(A, \omega), \quad A \in \mathscr{A}, \tag{4.4}$$

depending on $\omega \in \Omega_S$ and $S \in \mathscr{S}$ and having the consistency properties (4.2). For clarity we will assume that $\mathscr{S}$ consists of all domains $S \subseteq T$ which are complements of compacts in the parameter space T.

We consider an example which is applicable to the case when T is discrete.

EXAMPLE. Let $(X_t, \mathscr{B}_t)$, $t \in T$ be a collection of measurable spaces with measures Q_t. For each $S \subseteq T$ we let $X_S = \prod_{t \in S} X_t$, the coordinate space of elements $x_S = \{x(t), t \in S\}$, where $x(t) \in X_t$, and $Q_S = \prod_{t \in S} Q_t$ be the product measure on X_S with σ-algebra generated by the semi-ring of cylinder sets $A_S \in \prod_{t \in S} \mathscr{B}_t$. We can represent each cylinder set $A \in \prod_{t \in T} \mathscr{B}_t$ in the space $X = X_T$ in the form $A = A_{S_1} \times A_S$ for any $S \subseteq T$, $S_1 = T \backslash S$. By $\mathscr{A}(S)$ we mean the σ-algebra in X which is generated by all cylinder sets of the form $A = X_{S_1} \times A_S$. Let $p(x)$, $x \in X$, be a non-negative function measurable with respect to the σ-algebra $\mathscr{A} = A(T)$ such that for $S \in \mathscr{S}$ and all $x = (x_{S_1}, x_S) \in X$

$$\int_{X_{S_1}} p(x) Q_{S_1}(dx_{S_1}) < \infty.$$

Then analogous to (2.17) the equations

$$P_S(A_{S_1} \times A_S, x) = \int_{A_{S_1}} p_{S_1}(x) Q_{S_1}(dx_{S_1}) \times 1_{A_S}(x_S),$$

$$p_{S_1}(x) = \frac{p(x)}{\int_{X_{S_1}} p(x) Q_{S_1}(dx_{S_1})}, \qquad x = (x_{S_1}, x_S), \tag{4.5}$$

give a family $P_S(\cdot, x)$, $S \in \mathscr{S}$, of consistent conditional distributions on the semi-ring of cylinder sets $\prod_{t \in T} \mathscr{B}_t$ generating the σ-algebra $\mathscr{A}$ in the space $\Omega = X$. In the case where $p(x)$ is a probability density with respect to $Q = \prod_{t \in T} Q_t$ ($\int_X p(x) Q(dx) = 1$), these are consistent conditional distributions for the probability measure P,

$$P(A) = \int_A p(x) Q(dx), \qquad A \in \mathscr{A}.$$

EXAMPLE (Conditional Distributions of Markov Type). Let $\mathscr{S}$ be a collection of domains which are complements of compacts in the space T. We will suppose that the function $p(x)$ in (4.5) has the multiplicative structure shown in (2.18): for every $S_1 = T \backslash S, S \in \mathscr{S}$, one can find a compact set $S_2 \subseteq S$ such that

$$p(x) = p_{S_1}(x_{S_1}, x_{S_2}) p_{S_2}(x_{S_2}, x_{S_3}), \qquad S_3 = S \backslash S_2. \tag{4.6}$$

Then we can write (4.5) as

$$p_{S_1}(x) = \frac{p_{S_1}(x_{S_1}, x_{S_2})}{\int_{X_{S_1}} p_{S_1}(x_{S_1}, x_{S_2}) Q_{S_1}(dx_{S_1})}.$$

Clearly, in establishing the conditional distributions (4.5) it is not necessary to take an arbitrary function $p(x)$ defined on the whole space X; it is sufficient to assign, for every $S \in \mathscr{S}$, a corresponding factor $p_{S_1}(x_{S_1}, x_{S_2})$. More precisely, take a sequence of compacts $S_k', k = 1, 2, \ldots$, such that the union of these is the whole space T and for each $S \in \mathscr{S}$ there is some n for which the complement of the compact set $\bigcup_{k \leq n} S_k'$ is contained in S. We form the formal product

$$p(x) = \prod_k p_k(x_{S_k'}, x_{S_{k+1}'}).$$

With this definition of $p(x)$, for every $S \in \mathscr{S}$, equation (4.5) involves only a finite product of factors $p_k(x_{S_k'}, x_{S_{k+1}'}), k \leq n$, where n is defined by the inclusion $(\bigcup_{k \leq n} S_k')^C \subseteq S$. The consistent conditional distributions given by (4.5) have the property that for every $S \in \mathscr{S}$, one can find a compact set $S_2 \subseteq S$ such that (4.6) is formally true and for all $A \in \mathscr{A}(S_1)$ the conditional probabilities $P_S(A, x)$ are measurable with respect to the σ-algebra $\mathscr{A}(S_2)$. For the probability measure P with conditional distributions (4.5) we have

$$P_S(A, x) = P(A | \mathscr{A}(S)) = P(A | \mathscr{A}(S_2)), \qquad A \in \mathscr{A}(S_1). \tag{4.7}$$

Questions about the existence and uniqueness of a probability measure P for which these distributions are conditional distributions with respect to the corresponding σ-algebras $\mathscr{A}(S)$ are nontrivial, cf. §4.1.

Of course, if we restrict ourselves to considering only sets $S \supseteq S_0$, for some fixed $S_0 \in \mathscr{S}$, then these questions are answered very simply: each probability measure Q on the σ-algebra $\mathscr{A}(S_0)$ with support Ω_{S_0} gives the desired probability measure P by

$$P(A) = \int_\Omega P_{S_0}(A, \omega) Q(d\omega), \qquad A \in \mathscr{A}. \tag{4.8}$$

Actually what we have is that for $S \supseteq S_0$, the equality

$$P_{S_0}(AB, \omega') = \int_B P_S(A, \omega) P_{S_0}(d\omega, \omega')$$

holds for all $B \in \mathscr{A}(S)$, hence

$$P(AB) = \int_B P_S(A, \omega)P(d\omega), \qquad B \in \mathscr{A}(S), \qquad (4.9)$$

since

$$P(AB) = \int_\Omega P_{S_0}(AB, \omega')Q(d\omega') = \int_\Omega \int_B P_S(A, \omega)P_{S_0}(d\omega, \omega')Q(d\omega').$$

Clearly, if we complete the collection (4.4) with conditional distributions $P_{S_0}(\cdot, \omega)$ corresponding to $S_0 = \infty$ and defined on some set $\Omega_\infty \in \mathscr{A}(\infty)$, then formula (4.8) gives us the measure P we need so that (4.9) will be satisfied for all S.

Conversely, such a completion is always possible if there exists a probability measure P with given conditional distributions.

In fact, if we take the conditional distributions P_∞ described in (3.9), for all $A \in \mathscr{A}$ and $B \in \mathscr{A}(S)$, $S \subseteq \mathscr{S}$, we get

$$P_\infty(AB, \omega') = \int_B P_S(A, \omega)P_\infty(d\omega, \omega'), \qquad \omega' \in \Omega_\infty. \qquad (4.10)$$

This allows us to integrate the following expression with respect to the measure $P_\infty(\cdot, \omega')$:

$$P_{S_n}(AB, \omega'') = \int_B P_S(A, \omega)P_{S_n}(d\omega, \omega''), \qquad S_n \subseteq S;$$

we remark for clarity that the measure

$$\int_\Omega P_{S_n}(A, \omega'')P_\infty(d\omega'', \omega'), \qquad A \in \mathscr{A},$$

which appears on the right side after the integration coincides with the measure $P_\infty(A, \omega')$, $A \in \mathscr{A}$.

Recall that on the set Ω_∞ of full measure $P(\Omega_\infty) = 1$, the conditional distributions $P_\infty(\cdot, \omega)$ can be obtained from the original collection (4.4) by taking the limit

$$\lim P_{S_n}(A, \omega) = P_\infty(A, \omega), \qquad A \in \mathfrak{G}(\mathscr{A}), \qquad (4.11)$$

where $\mathfrak{G}(\mathscr{A})$ is a countable algebra generating the σ-algebra $\mathscr{A}$.

It is clear that the conditional distributions $P_\infty(\cdot, \omega)$ which complete the original collection (4.4) are among the distributions obtained from the limit (4.11). Namely, we define a set $\Omega_\infty^0 \in \mathscr{A}(\infty)$ as the collection of all ω for which the limit (4.11) exists for all $A \in \mathfrak{G}(\mathscr{A})$; moreover, the limit $P_\infty(A, \omega)$ is a distribution on the algebra $\mathfrak{G}(\mathscr{A})$ and thus it extends to a measure on the σ-algebra $\mathscr{A}$. Then for any probability measure P with conditional distributions (4.4) we have $\Omega_\infty \subset \Omega_\infty^0$ for a corresponding set Ω_∞ and

$$P(A) = \int_\Omega P_\infty(A, \omega)P(d\omega). \qquad (4.12)$$

We can think of P as belonging to the convex hull of the limiting distributions $P_\infty(\cdot, \omega)$, almost all of which are conditional distributions for P. The question of existence and uniqueness of a probability measure P with the given conditional distributions (4.4) leads to the problem of identifying such a measure among the limiting distributions (4.11).

We consider an important example where each limit distribution (4.11) has the given family (4.4) as its conditional distributions.

Let the space X be finite and the parameter space T be discrete. Let all the probabilities $P_S(A, \omega)$, $A \in \mathcal{A}$, in the given consistent collection (4.4) be measurable with respect to the corresponding algebras

$$\mathcal{A}_0(S) = \bigcup_{S' \subseteq S} \mathcal{A}(S')$$

where the union is taken over all compact $S' \subseteq S$. For the countable algebra $\mathfrak{G}(\mathcal{A})$ in the definition of the limit distributions (4.11) we can take $\mathfrak{G}(\mathcal{A}) = \mathcal{A}_0(T)$, where

$$\mathcal{A}_0(T) = \bigcup_{S \subseteq T} \mathcal{A}(S)$$

is the union over all compact $S \subseteq T$. (There are not more than a countable number of such S and each of the σ-algebras $\mathcal{A}(S)$ is finite.) Every real function $\varphi(\omega)$ which is measurable with respect to the algebra $\mathcal{A}_0(T)$ is the uniform limit of some sequence of step functions of the form

$$\sum c_k 1_{A_k}, \qquad A_k \in \mathfrak{G}(\mathcal{A}),$$

where for each $A = A_k$

$$\lim P_{S_n}(A, \omega') = P_\infty(A, \omega'), \qquad \omega' \in \Omega_\infty^0,$$

and consequently, for each $B \in \mathcal{A}_0(T)$

$$\lim \int_B \varphi(\omega) P_{S_n}(d\omega, \omega') = \int_B \varphi(\omega) P_\infty(d\omega, \omega').$$

For $\varphi(\omega) = P_S(A, \omega)$ and $B \in \mathcal{A}_0(S)$ we get

$$\int_B P_S(A, \omega) P_\infty(d\omega, \omega') = \lim \int_B P_S(A, \omega) P_{S_n}(d\omega, \omega')$$

$$= \lim P_{S_n}(AB, \omega') = P_\infty(AB, \omega').$$

It is evident that the consistent family (4.4) gives conditional distributions for each probability measure $P_\infty(\cdot, \omega)$, which we define by taking the limit (4.11).

EXAMPLE. Let $T = \mathbb{Z}^1$ be the set of integers $t = 0, \pm 1, \ldots$, and $\xi = \xi(t)$, $t \geq t_0$ be a homogeneous aperiodic Markov chain with a finite number of states $x \in X$ and varying initial parameter $t_0 \in T, t_0 \to -\infty$. In the coordinate space $\Omega = X^T$ of sequences $\omega = \{x(t), t \in T\}$ with values in X, we define consistent conditional distributions (4.4), specifying them on the semi-ring of cylinder sets of the form

$$A = A_1 \times A_2; \quad A_1 \in \mathcal{A}(S_1), \quad A_2 \in \mathcal{A}(S),$$

as follows. For $S \subseteq T$ the complement of the interval $S_1 = \{u < t < v\}$ we let

$$P_S(A, \omega) = P\{\xi \in A_1 \,|\, \xi(u) = x(u), \xi(v) = x(v)\} \cdot 1_{A_2}(\omega)$$

and for any $S \supseteq S_0$, where $S_0 \subseteq T$ is a set of the above type, we define $P_S(A, \omega)$ as the conditional distributions for the probability measure $P_{S_0}(\cdot, \omega')$. We consider the sequence of complements S of increasing intervals (u, v), $u \to -\infty$, $v \to +\infty$. Clearly, the limit

$$\lim P_S(A, \omega) = P_\infty(A, \omega)$$

gives us a probability distribution $P_\infty(\cdot, \omega)$ if and only if every coordinate $x(t)$ of the point $\omega = \{x(t), t \in T\}$ belongs to the same ergodic class, and furthermore

$$P_\infty(A, \omega) = P(A, x)$$

is the stationary probability of the corresponding ergodic class with representative $x \in X$.

3. Construction of Consistent Conditional Distributions†

We will consider a system of σ-algebras $\mathcal{A}_t \subseteq \mathcal{A}$ in a probability space $(\Omega, \mathcal{A}, P)$, depending on a parameter t which runs through a partially ordered set $\mathcal{T}$ with order relation $\leq$. We will assume that

$$\mathcal{A}_s \subseteq \mathcal{A}_t \quad \text{if } s \leq t. \tag{4.13}$$

In the previous discussion we considered the special case where $\mathcal{T} = \mathcal{S}$ was a collection of subsets $S \subseteq T$ ordered by inclusion.

For every $t \in \mathcal{T}$ let there be given the conditional distribution

$$P_t(A, \omega) = P\{A \,|\, \mathcal{A}_t\}, \quad A \in \mathcal{A}, \tag{4.14}$$

defined for all $\omega \in \Omega$. We say that the conditional distributions (4.14) are *consistent* if for $t \geq s$

$$P_t(A, \omega) = P_s(A \,|\, \mathcal{A}_t), \quad A \in \mathcal{A} \tag{4.15}$$

are conditional probabilities for each probability measure $P_s(\cdot, \omega')$, $\omega' \in \Omega$.

We remark that (4.15) is a slightly stronger requirement than condition (4.2), in which the consistency condition was required for ω' only in some set of full measure.

---

† This section was written by C. E. Kuznetzov.

Let the probability measure P be perfect and let the σ-algebra $\mathcal{A}$ and all the σ-algebras $\mathcal{A}_t$, $t \in \mathcal{T}$ be separable and suppose we have:

for every $s \in \mathcal{T}$, the set $\{t: t \in \mathcal{T}, t > s\}$ is countable.   (4.16)

Under these conditions, consistent conditional distributions exist and we will show this. It was shown in §2 that in our given framework of assumptions conditional distributions

$$P_t(A, \omega) = P(A \mid \mathcal{A}_t), \qquad A \in \mathcal{A},   \tag{4.17}$$

defined for all $\omega \in \Omega$ exist. In view of (4.16) and (2.22), for every $s \in \mathcal{T}$ there is a set $\Omega_s \in \mathcal{A}_s$ with $P(\Omega_s) = 1$, such that for $t > s$

$$P_t(A, \omega) = P_s(A \mid \mathcal{A}_t), \qquad A \in \mathcal{A},$$

for any measure $P_s(\cdot, \omega')$, $\omega' \in \Omega_s$.

We construct for each $s \in \mathcal{T}$ a monotonic sequence of sets Ω_s^n, with $\Omega_s^1 = \Omega_s$, and having the property that the intersection of Ω_s^n with Ω_s^{n-1} is the set

$$\{\omega: P_s(\Omega_t^{n-1}, \omega) = 1 \text{ for all } t > s\}.$$

From (4.16), $\Omega_s^n \in \mathcal{A}_s$ and $P(\Omega_s^n) = 1$. Let $\Omega_s^\infty = \bigcap_n \Omega_s^n$. Clearly,

$$\Omega_s^\infty \in \mathcal{A}_s \quad \text{and} \quad P(\Omega_s^\infty) = 1,   \tag{4.18}$$

$$P_s(\Omega_t^\infty, \omega) = 1, \quad \text{for } s < t, \, \omega \in \Omega_s^\infty.   \tag{4.19}$$

We define the desired conditional distributions by the formula

$$\tilde{P}_s(A, \omega) = \begin{cases} P_s(A, \omega), & \text{if } \omega \in \Omega_s^\infty, \\ P(A), & \text{otherwise.} \end{cases}   \tag{4.20}$$

In view of (4.17) and (4.18), $\tilde{P}_s$ are conditional distributions. To prove their consistency we must show that for each probability measure $\tilde{P}_s(\cdot, \omega')$, $\omega' \in \Omega$, and for any $t \geq s$,

$$\tilde{P}_t(A, \omega) = \tilde{P}_s(A \mid \mathcal{A}_t), \qquad A \in \mathcal{A}.$$

First let $\omega' \in \Omega_s^\infty$. Then from the definition of Ω_s^∞ and (4.19)–(4.20),

$$\tilde{P}_s(A \mid \mathcal{A}_t) = P_s(A \mid \mathcal{A}_t) = P_t(A, \omega) = \tilde{P}_t(A, \omega)$$

almost everywhere under the probability measure $\tilde{P}_s(\cdot, \omega') = P_s(\cdot, \omega')$. For $\omega' \notin \Omega_s^\infty$, just apply (4.17)–(4.18).

Using these results, we can construct consistent distributions without the countability condition (4.16), provided there is a subset $\mathcal{T}_0 \subset \mathcal{T}$ satisfying (4.16) and such that any σ-algebra $\mathcal{A}_t$, $t \in \mathcal{T}$, can be approximated by a monotonic sequence of σ-algebras $\mathcal{A}_{t_n}$, $t_n \in \mathcal{T}_0$. We will provide the necessary details. Let there be a collection of σ-algebras satisfying (4.13) and a subset $\mathcal{T}_0 \subset \mathcal{T}$ such that for any $s \in \mathcal{T}$

$$\{t: t \in \mathcal{T}_0, t > s\} \text{ is a countable set.}   \tag{4.21}$$

We call $\mathcal{T}_0$ an *exterior skeleton* for $\mathcal{A}_t$, $t \in \mathcal{T}$, if for any $t \in \mathcal{T} \setminus \mathcal{T}_0$ we can find a sequence $t_n \in \mathcal{T}_0$ such that

$$t_1 \geq t_2 \geq \cdots \geq t_n \geq \cdots \geq t, \tag{4.22}$$

$$\mathcal{A}_t = \bigcap_n \mathcal{A}_{t_n}. \tag{4.23}$$

We call $\mathcal{T}_0$ an *interior skeleton* if for any $t \in \mathcal{T}$ we can find a sequence $t_n \in \mathcal{T}_0$ such that

$$t_1 \leq t_2 \leq \cdots \leq t_n \leq \cdots \leq t, \tag{4.24}$$

and

$$\mathcal{A}_t = \bigvee_n \mathcal{A}_{t_n}, \tag{4.25}$$

and for any $s < t$ there is some N such that

$$s < t_N < t. \tag{4.26}$$

Theorem. *Let the system of σ-algebras $\mathcal{A}_t$, $t \in \mathcal{T}$ have an (internal or external) skeleton $\mathcal{T}_0$, let the σ-algebras $\mathcal{A}_t$, $t \in \mathcal{T}_0$, be separable and the σ-algebra $\mathcal{A}$ in the probability space $(\Omega, \mathcal{A}, P)$ be a standard Borel σ-algebra. Then consistent conditional distributions exist.*

We will show this. Note that a standard Borel σ-algebra is separable and any probability measure on it is perfect. It has already been shown that for the system of σ-algebras $\mathcal{A}_t$, $t \in \mathcal{T}_0$, there exist consistent conditional distributions $P_t(\cdot, \omega) = P(\cdot|\mathcal{A}_t)$, $t \in \mathcal{T}_0$. Let $s \in \mathcal{T} \setminus \mathcal{T}_0$. We will look at the collection $\mathcal{P}_s$ of all probability measures P' on the σ-algebra $\mathcal{A}$ satisfying the condition: for any $t > s$, $t \in \mathcal{T}_0$,

$$P_t(A, \omega) = P'(A|\mathcal{A}_t). \tag{4.27}$$

We remark that the set $\mathcal{P}_s$ contains the measure P and all measures $P_{s'}(\cdot, \omega)$ for $s' \in \mathcal{T}_0$ with $s' < s$. Thus it is clear that

$$\mathcal{P}_s \subset \mathcal{P}_t, \qquad s < t. \tag{4.28}$$

The following lemma is used in the construction of the desired conditional probabilities.

Lemma. (a) *Let $\mathcal{T}_0$ be an exterior skeleton. For each $t \in \mathcal{T} \setminus \mathcal{T}_0$ there are functions $P_t(A, \omega)$, $A \in \mathcal{A}$, $\omega \in \Omega$, which are the common conditional distributions with respect to $\mathcal{A}_t$ for all measures $P' \in \mathcal{P}_t$:*

$$P_t(A, \omega) = P'(A|\mathcal{A}_t), \qquad A \in \mathcal{A}. \tag{4.29}$$

In addition, for every $\omega \in \Omega$ the measure $P_t(\cdot, w)$ belongs to the class $\mathcal{P}_t$:

$$P_t(\cdot, \omega) \in \mathcal{P}_t, \qquad \omega \in \Omega. \tag{4.30}$$

(b) *Let $\mathcal{T}_0$ be an interior skeleton. For every $t \in \mathcal{T}$ there are functions $P_t(A, \omega)$, $A \in \mathcal{A}$, $\omega \in \Omega$, which are the common conditional distributions with respect to $\mathcal{A}_t$ for all measures $P' \in \mathcal{P}_s$, $s < t$:*

$$P_t(A, \omega) = P'(A \mid \mathcal{A}_t), \qquad A \in \mathcal{A}. \tag{4.31}$$

In addition, for every $\omega \in \Omega$ the measure $P_t(\cdot, \omega)$ belongs to $\mathcal{P}_t$:

$$P_t(\cdot, \omega) \in \mathcal{P}_t. \tag{4.32}$$

First let us see how this lemma leads to the proof of the theorem. For instance, let $\mathcal{T}_0$ be an exterior skeleton. Conditional distributions were defined already for $t \in \mathcal{T}_0$. For the remaining t we will use the distributions whose existence is guaranteed by the lemma. The fact that $P \in \mathcal{P}_t$ and condition (4.29) give us (4.14). We will show that the consistency condition (4.15) is satisfied. In fact, it was already checked for $s, t \in \mathcal{T}_0$. When $s \notin \mathcal{T}_0$, $t \in \mathcal{T}_0$, then it follows from (4.30) and (4.27). If $s \in \mathcal{T}_0$ and $t \notin \mathcal{T}_0$, then all the measures $P_s(\cdot, \omega)$, $\omega \in \Omega$, belong to $\mathcal{P}_t$ and the conclusion follows from (4.29). Finally, if s and t belong to $\mathcal{T} \setminus \mathcal{T}_0$, then the measures $P_s(\cdot, \omega)$, $\omega \in \Omega$, belong to $\mathcal{P}_s$ and consequently $\mathcal{P}_t$ in view of (4.30) and (4.28); the rest follows from (4.29). The case when $\mathcal{T}_0$ is an interior skeleton proceeds in a similar fashion.

PROOF OF THE LEMMA. First let $\mathcal{T}_0$ be an exterior skeleton. Without loss of generality we can assume that Ω is compact metric and $\mathcal{A}$ is the Borel σ-algebra on Ω. Let $C(\Omega)$ be the space of continuous functions on Ω and W be a countable dense subset of $C(\Omega)$. Let P_n be probability measures on the σ-algebra A such that the sequence of functionals

$$l_n(f) = \int f \, dP_n$$

converges for all $f \in W$. Then the sequence $l_n(f)$ converges for all $f \in C(\Omega)$, since

$$\left| \int f \, dP_n - \int g \, dP_n \right| \le \sup_\Omega |f - g|$$

for any $f \in C(\Omega)$, $g \in W$. Therefore the sequence of measures P_n converges weakly to a probability measure P; that is,

$$\lim \int f \, dP_n = \int f \, dP, \qquad f \in C(\Omega).$$

We employ this argument for the measures $P_n = P_{t_n}(\cdot, \omega)$, where t_n are points in the exterior skeleton $\mathcal{T}_0$ (cf. (4.22)–(4.23)). We consider the functions

$$l_n(f, \omega') = \int_\Omega f(\omega) P_{t_n}(d\omega, \omega')$$

and denote by Ω_t the set of all ω' for which the sequence $l_n(f, \omega')$ converges for all $f \in W$. For each ω' in Ω_t there exists a unique probability measure $P_t(\cdot, \omega')$ on the σ-algebra $\mathscr{A}$ such that

$$\int f(\omega) P_t(d\omega, \omega') = \lim_n l_n(f, \omega'), \qquad f \in C(\Omega). \tag{4.33}$$

Since W is countable and the function $l_n(f, \omega')$ is measurable with respect to $\mathscr{A}_{t_n}$ for every f and n, $\Omega_t \in \mathscr{A}_t$ and the left side of (4.33) is an $\mathscr{A}_t$-measurable function. Furthermore, for any $P' \in \mathscr{P}_t$,

$$l_n(f, \omega') = E'(f | \mathscr{A}_{t_n}),$$

where $E'(\cdot | \cdot)$ is the conditional expectation for the measure P', and thus $P'(\Omega t) = 1$ for all $P' \in \mathscr{P}_t$ and

$$\int f(\omega) P_t(d\omega, \omega') = E'(f | \mathscr{A}_t) \tag{4.34}$$

for all $P' \in \mathscr{P}_t$ and all $f \in C(\Omega)$. Since the conditional expectation $E'(\cdot | \mathscr{A}_t)$ is a continuous linear operator on L^1, (4.34) is true for all bounded functions f which are pointwise limits of bounded sequences of functions in $C(\Omega)$ and for linear combinations of such functions. In particular (4.34) holds for all indicator functions $f = 1_A$, where A is in the semi-ring $\mathfrak{G}$ of sets of the form $A = G_1 \backslash G_2$, G_1 and G_2 open. Now for functions $f = 1_A$, $A \in \mathscr{A}$, (4.34) can be rewritten in the form

$$\int_B P_t(A, \omega') P'(d\omega') = P'(AB) \tag{4.35}$$

for all $B \in \mathscr{A}_t$. For each fixed B, both sides of (4.35), when regarded as functions of A, are measures which coincide on the semi-ring $\mathfrak{G}$ which generates the Borel σ-algebra $\mathscr{A}$. Consequently, (4.35) is true for all $A \in \mathscr{A}$ and $P_t(\cdot, \omega) = P'(\cdot | \mathscr{A}_t)$ is the common conditional distribution for all measures $P' \in \mathscr{P}_t$ with respect to the σ-algebra $\mathscr{A}_t$.

We let $\omega' \in \tilde{\Omega}_t$ if the measure $P_t(\cdot, \omega')$ belongs to $\mathscr{P}_t$. We define the desired conditional distribution by

$$\tilde{P}_t(A, \omega) = \begin{cases} P_t(A, \omega), & \text{if } \omega \in \tilde{\Omega}_t, \\ P(A), & \text{otherwise.} \end{cases}$$

Clearly the measures $\tilde{P}_t(\cdot, \omega)$ belong to $\mathscr{P}_t$ for all ω. The proof of the lemma will be complete if we show that

$$\tilde{\Omega}_t \in \mathscr{A}_t \quad \text{and} \quad P'(\tilde{\Omega}_t) = 1 \tag{4.36}$$

for all $P' \in \mathscr{P}_t$. In fact, in that case the measures $\tilde{P}_t(\cdot, \omega)$ and $P_t(\cdot, \omega)$ coincide for P'-almost all ω, for all $P' \in \mathscr{P}_t$. Thus $\tilde{P}_t(\cdot, \omega)$, as well as $P_t(\cdot, \omega)$, are common conditional distributions for all measures P' in the class $\mathscr{P}_t$.

For the proof of (4.36) we make use of the separability of the σ-algebra $\mathscr{A}$ and all the σ-algebras $\mathscr{A}_t$, $t \in \mathscr{T}_0$. Let $\mathfrak{G} = \mathfrak{G}(\mathscr{A})$, $\mathfrak{G}_t = \mathfrak{G}(\mathscr{A}_t)$ be countable algebras generating the corresponding σ-algebras. Comparing the definition of $\mathscr{P}_t$ (cf. (4.27)) with the definition of $\tilde{\Omega}_t$, we see that $\omega \in \tilde{\Omega}_t$ if and only if

$$\int_B P_{t'}(A, \omega') P_t(d\omega', \omega) = P_t(AB, \omega) \tag{4.37}$$

for all $t' > t, t' \in \mathscr{T}_0, A \in \mathfrak{G}, B \in \mathfrak{G}_{t'}$. For each fixed A, B, t', both sides of (4.37) are $\mathscr{A}_t$ measurable. Consequently the set $\tilde{\Omega}_t$ is also $\mathscr{A}_t$-measurable. By (4.21) all that is left to show is that for any fixed A, B, t', and any measure $P' \in \mathscr{P}_t$, equation (4.37) holds for P'-almost all ω. We already showed that $P_t(\cdot, \omega) = P'(\cdot | \mathscr{A}_t)$, hence

$$\begin{aligned}
P_t(AB, \omega) &= P'(AB | \mathscr{A}_t) \\
&= E'(P'(AB | \mathscr{A}_{t'}) | \mathscr{A}_t) \\
&= E'(1_B P'(A | \mathscr{A}_{t'}) | \mathscr{A}_t) \\
&= E'(1_B(\cdot) P_{t'}(A, \cdot) | \mathscr{A}_t) \\
&= \int 1_B(\omega') P_{t'}(A, \omega') P_t(d\omega', \omega) \\
&= \int_B P_{t'}(A, \omega') P_t(d\omega', \omega).
\end{aligned}$$

The proof for the case when $\mathscr{T}_0$ is an interior skeleton is carried out in the same way. $\square$

EXAMPLE (System of σ-algebras Generated by Generalized Random Functions). Let $C_0^\infty = C_0^\infty(\mathbb{R}^d)$ be the space of infinitely differentiable functions $u(t), t \in \mathbb{R}^d$, with compact support Supp u. We look at the space $X = (C_0^\infty)^*$ of generalized functions, that is, all continuous linear functionals $x = \langle x, u \rangle$ on the space C_0^∞. We define $\mathscr{B}$ to be the σ-algebra in the space X which is generated by all functions $F(x), x \in X$, of the form

$$F(x) = \langle x, u \rangle, \qquad u \in C_0^\infty.$$

Let $\xi = \xi(\omega), \omega \in \Omega$, be a measurable mapping from a probability space $(\Omega, \mathscr{A}, P)$ into the space $(X, \mathscr{B})$. Such a mapping is a random continuous linear functional (see §5.3 in this chapter) and will be called a *generalized random function*. Under certain conditions, such as ξ Gaussian, ξ is a generalized random function in the sense of §1.3. Each generalized random function induces a probability measure P_ξ on the σ-algebra $\mathscr{B}$ in the space X:

$$P_\xi(B) = P\{\xi(\omega) \in B\}, \qquad B \in \mathscr{B}.$$

Turning to the measure P_ξ, we can assume without loss of generality that $\Omega = X$ and $\mathscr{A} = \mathscr{B}$ and the mapping $\xi: \Omega \to X$ is the identity.

Let S be an open domain in $\mathbb{R}^d$. There is a σ-algebra $\mathscr{A}(A)$ connected with S which is generated by all functions of the form

$$\xi_u(\omega) = \langle \xi(\omega), u \rangle, \qquad \text{Supp } u \subseteq S.$$

When S is closed we let

$$\mathscr{A}(S) = \bigcap_{\varepsilon > 0} \mathscr{A}(S^\varepsilon).$$

We denote by $\mathscr{T}_1$ the collection of all closed sets $S \subseteq \mathbb{R}^d$ which are bounded or whose complements are bounded. We introduce the natural partial ordering by set inclusion $\subseteq$ on $\mathscr{T}_1$. Let $\mathscr{T}_2$ be the collection of all open sets $S \subseteq \mathbb{R}^d$ which are bounded or have bounded complements. We introduce on $\mathscr{T}_2$ a partial ordering by *strong inclusion*:

$$S_1 \leq S_2 \quad \text{if } \bar{S}_1 \subseteq S_2.$$

We will show that for each of the systems of σ-algebras, $\mathscr{A}(S)$, $S \in \mathscr{T}_1$ and $\mathscr{A}(S)$, $S \in \mathscr{T}_2$, consistent conditional distributions exist.

Note that when S is open, the σ-algebra $\mathscr{A}(S)$ is separable since then it is generated by a countable collection of random functions ξ_{u_n}, $n = 1, 2, \ldots$, where the u_n are a countable dense subset of $C_0^\infty(S)$.

We will construct a special system of countably many open sets so that there is a collection $\mathscr{T}_0$ of sets which form an exterior skeleton for $\mathscr{A}(S)$, $S \in \mathscr{T}_1 \cup \mathscr{T}_0$, and an interior skeleton for $\mathscr{A}(S)$, $S \in \mathscr{T}_2$. Let $\mathfrak{M}_k$, $k = 0, 1, 2, \ldots$, be the collection of all closed sets $A \subset \mathbb{R}^d$ of the form

$$A = \{t = (t_1, \ldots, t_d) : n_i 2^{-k} \leq t_i \leq (n_i + 1)2^{-k}, i = 1, \ldots, d\} \quad (4.38)$$

where n_i are integers. $\mathscr{T}_0$ consists of any set which is either the complement of or the interior of a finite sum of sets in $\mathfrak{M} = \bigcup_k \mathfrak{M}_k$.

Let S be a closed bounded set. We denote by S_k the interior of the union of all sets in $\mathfrak{M}_k$ having nonempty intersection with S. Note that the open set S_k contains the closed set S. In fact, if this were not so then there would be a point $t \in S$ belonging to the boundary ∂S_k of the set S_k. But then it would not be hard to find a set of the form (4.38), containing the point t and not contained in $\bar{S}_k$, which contradicts the definition of S_k. The following properties hold:

$$S_k \supseteq S_{k+1}; \quad (4.39)$$

for any $\varepsilon > 0$ there is an ε-neighborhood S^ε of the set S and for sufficiently large k

$$S_k \subseteq S^\varepsilon; \quad (4.40)$$

for any k and sufficiently small $\delta > 0$

$$S^\delta \subseteq S_k. \quad (4.41)$$

The inclusion (4.40) follows from the fact that the diameters of sets in $\mathfrak{M}_k$ approach zero as $k \to \infty$, and the inclusion (4.41) is a consequence of the compactness of S.

Similarly, for a closed set S with bounded complement we denote by S_k the complement of the union of all sets in $\mathfrak{M}_k$ which are entirely contained in S^c. Clearly $S_k \supseteq S$ and (4.39)–(4.41) are true. From (4.40)–(4.41)

$$\mathscr{A}(S) = \bigcap_k \mathscr{A}(S_k).$$

Therefore the sequence S_k satisfies (4.22)–(4.23) and $\mathscr{T}_0$ is an exterior skeleton for $\mathscr{A}(S)$, $S \in \mathscr{T}_1 \cup \mathscr{T}_0$.

We will show that $\mathscr{T}_0$ is also an interior skeleton for $\mathscr{A}(S)$, $S \in \mathscr{T}_2$. Let $S \in \mathscr{T}_2$ be an open set. Then $S^c \in \mathscr{T}_1$ and there exists a sequence S'_k of elements of $\mathscr{T}_0$ approximating S^c in the sense of (4.39)–(4.41). We define S_k as the complement of $\bar{S}'_k$. Clearly $S_k \in \mathscr{T}_0$. By (4.39)–(4.41)

$$S_k \subseteq S_{k+1} \subseteq \cdots \subseteq S$$

and for any $\varepsilon > 0$ there is a sufficiently large k such that

$$S_k \supseteq S^{-\varepsilon} = \mathbb{R}^d \backslash (S^c)^\varepsilon. \tag{4.42}$$

In view of (4.42) $S = \bigcup_k S_k$ and in accordance with the definition of $\mathscr{A}(S)$,

$$\mathscr{A}(S) = \bigvee_k \mathscr{A}(S_k).$$

Thus the sequence S_k satisfies conditions (4.24)–(4.25). In addition, if S', $S \in \mathscr{T}_2$ and $\bar{S}' \subseteq S$, then $S' \subseteq S^{-\varepsilon}$ for some $\varepsilon > 0$, so that (4.26) follows from (4.42). Hence $\mathscr{T}_0$ is an interior skeleton for $\mathscr{A}(S)$, $S \in \mathscr{T}_2$.

In order to use the theorem proved earlier, it remains to show that the σ-algebra $\mathscr{B}$ in the space X is a standard Borel σ-algebra.

We show this by constructing a measurable map α from the space X into a complete separable metric space Y (the Hilbert cube) and verifying that $\alpha(X)$ is a Borel subset of Y and α is an isomorphism between X and $\alpha(X)$, i.e., there is a one-to-one correspondence between both sides.

The Hilbert cube Y is a countable product of intervals $[-1, 1]$, where each point $y = (y_1, y_2, \ldots)$ is defined by coordinates $y_k \in [-1, 1]$ and the distance between $y', y'' \in Y$ is given by

$$\rho(y', y'') = \sum_k \frac{1}{2^k} |y'_k - y''_k|.$$

In order to construct the map α, we choose a countable dense family of functions $\mathscr{K}$ in C_0^∞ such that for any ball

$$S_r = \{x : |x| < r\} \subseteq \mathbb{R}^d, \qquad r = 1, 2, \ldots,$$

the collection $\mathscr{K}_r$ of functions in $\mathscr{K}$ with support in S_r is dense in $C_0^\infty(S_r)$ and such that $c_1 u_1 + c_2 u_2 \in \mathscr{K}$ if $u_i \in \mathscr{K}$ and c_i are rational. To get such a family of functions $\mathscr{K}$, it suffices to choose a countable dense set from each separable Hilbert space $C_0^\infty(S_r)$ and let $\mathscr{K}$ be all linear combinations (with rational coefficients) of functions in the union of these sets.

We define the map $\alpha: X \to Y$ by associating the generalized function $x = \langle x, u \rangle \in X$ with the point $\alpha(x) \in Y$ with coordinates

$$\alpha_k(x) = \frac{\langle x, u_k \rangle}{1 + |\langle x, u_k \rangle|}, \qquad u_k \in \mathcal{K}.$$

Each of the coordinates $\alpha_k(x)$ is a measurable function and thus the map $\alpha(x)$ is measurable. Note that

$$\langle x, u_k \rangle = \frac{\alpha_k(x)}{1 - |\alpha_k(x)|}. \tag{4.43}$$

For this reason if $x_1 \neq x_2$ then $\alpha(x_1) \neq \alpha(x_2)$. Furthermore, the σ-algebra $\mathcal{B}$ in the space X is generated by a countable number of functions $F(x) = \langle x, u_k \rangle$, $u_k \in \mathcal{K}$; hence the σ-algebra $\mathcal{B}$ is generated by the map α and any event $B \in \mathcal{B}$ has the form

$$B = \alpha^{-1}(C), \qquad C \in \mathcal{C}, \tag{4.44}$$

where $\mathcal{C}$ is the Borel σ-algebra in Y.

Now we describe the image $\alpha(X)$ of the space X. Let $\tilde{Y}$ be the set of points $y \in Y$ satisfying

$$|y_k| < 1, \qquad k = 1, 2, \ldots, \tag{4.45}$$

$$\frac{c_1 y_n}{1 - |y_n|} + \frac{c_2 y_m}{1 - |y_m|} = \frac{y_k}{1 - |y_k|} \tag{4.46}$$

for all n, m, k such that for c_1, c_2 rational, $c_1 u_n + c_2 u_m = u_k$. The set $\tilde{Y}$ is obviously Borel and $\alpha(X) \subset \tilde{Y}$ by (4.43).

Let p, q, r be natural numbers and let Y_{pqr} be the set of all points $y \in Y$ which satisfy, for all n, m such that $u_n, u_m \in \mathcal{K}_r$,

$$\left| \frac{y_n}{1 - |y_n|} - \frac{y_m}{1 - |y_m|} \right| \leq p \|u_n - u_m\|_q,$$

where $\|\cdot\|_q$ was defined by (1.12). It is clear that Y_{pqr} is a Borel set. We will show that $\alpha(X) = Y_0 = \tilde{Y} \cap (\bigcap_r \bigcup_{p,q} Y_{pqr})$.

Let $x \in X$. For every r, the functional x is continuous on the separable Hilbert space $C_0^\infty(S_r)$. Thus the functional x (as a functional on $C_0^\infty(S_r)$) is continuous in some norm $\|\cdot\|_q$. For sufficiently large p

$$|\langle x, u \rangle| \leq p \|u\|_q, \qquad u \in C_0^\infty(S_r).$$

By (4.43) for some r we can find p, q such that $\alpha(x) \in Y_{pqr}$. Thus $\alpha(x) \in Y_0$.

Conversely suppose y lies in Y_0. We define a functional $\langle x, u \rangle$ for a function $u \in \mathcal{K}$ by the equality

$$\langle x, u_k \rangle = \frac{y_k}{1 - |y_k|}.$$

For each r there are p, q such that $y \in Y_{pqr}$. Thus the functional $\langle x, u \rangle$ can be extended by continuity from $\mathcal{H}_r$ to $C_0^\infty(S_r)$ and from $\mathcal{H}$ to C_0^∞. Using (4.46) and (4.43), it is not hard to show that the resulting continuous functional is linear.

Thus $\alpha(X) = Y_0$ and we have shown that $\alpha(X)$ is Borel. Earlier we saw that the map α was one-to-one and measurable. The measurability of the inverse map α^{-1} easily follows from the measurability of $\alpha(X)$, since for a set B of the form (4.44)

$$\alpha(B) = \alpha(\alpha^{-1}(C)) = C \cap \alpha(X) \in \mathscr{C}.$$

§5. Gaussian Probability Distributions

1. Basic Definitions and Examples

A family of real random variables $\xi(t)$, $t \in T$, on a probability space $(\Omega, \mathscr{A}, P)$, where T is an arbitrary set, is called *Gaussian* if any choice of a finite number of the random variables $\xi(t_1), \ldots, \xi(t_n)$ has a Gaussian joint distribution, in other words, if the finite dimensional distributions of the random function $\xi(t)$, $t \in T$, are Gaussian. The random function, as well as the corresponding probability measure P on the σ-algebra $\mathscr{A}(T)$ generated by the variables $\xi(t)$, $t \in T$, are also called *Gaussian*.

Recall that a probability distribution P on the n-dimensional vector space $\mathbb{R}^n$ is *Gaussian* if the characteristic function

$$\varphi(u) = \int_{\mathbb{R}^n} \exp\{i(u, x)\} P(dx), \qquad u \in \mathbb{R}^n \tag{5.1}$$

has the form

$$\varphi(u) = \exp\{i(u, A) - \tfrac{1}{2}(Bu, u)\}, \qquad u \in \mathbb{R}^n;$$

here $(u, x) = \sum_{k=1}^n u_k x_k$ denotes the scalar product of the vectors $u = (u_1, \ldots, u_n)$ and $x = (x_1, \ldots, x_n)$, $A = (A_1, \ldots, A_n) \in \mathbb{R}^n$ is the "*mean value,*"

$$(u, A) = \int_{\mathbb{R}^n} (u, x) P(dx),$$

and B is the *covariance operator*—a positive linear operator given by the covariance matrix B_{kj}—with

$$(Bu, v) = \int_{\mathbb{R}^n} [(u, x) - (u, A)][(v, x) - (v, A)] P(dx), \qquad u, v \in \mathbb{R}^n.$$

If P is the joint distribution of Gaussian random variables $\xi_1, \ldots, \xi_n$ then

$$A_k = E\xi_k, \qquad B_{kj} = E(\xi_k - A_k)(\xi_j - A_j), \quad \text{for } k, j = 1, \ldots, n.$$

The probability distribution P with mean value A and covariance operator B is concentrated on an m-dimensional hyperplane L in the space $\mathbb{R}^n$ (m is the rank of the covariance matrix), which can be described by

$$L = A + B\mathbb{R}^n,$$

i.e., L is the collection of all vectors $y \in \mathbb{R}^n$ of the form $y = A + Bx$ for some $x \in \mathbb{R}^n$. In particular

$$P(\mathbb{R}^n \backslash L) = 0$$

and the distribution on L has a density function $p(y)$, $y \in L$, of the form

$$p(y) = \frac{1}{(2\pi)^{m/2}(\det B)^{1/2}} \exp\{-\tfrac{1}{2}(B^{-1}(y - A), y - A)\}. \qquad (5.2)$$

Here $\det B$ refers to the determinant of the matrix given by the operator B on the subspace $\mathbb{R}^m = B\mathbb{R}^n$ and B^{-1} is the operator on that subspace which is the inverse of $B|_{\mathbb{R}^m}$.

Each of the finite dimensional distributions $P_{t_1, \ldots, t_n}$ of the Gaussian random function $\xi = \xi(t)$ of $t \in T$ has mean value $[A(t_1), \ldots, A(t_n)]$ and covariance matrix $\{B(t_k, t_j)\}$, where $A(t)$, $t \in T$, is the *mean value* of the random function $\xi = \xi(t)$ and $B(s, t)$, $s, t \in T$, is its *covariance function*:

$$A(t) = E\xi(t)$$

$$B(s, t) = E[\xi(s) - A(s)][\xi(t) - A(t)]; \qquad s, t \in T.$$

Thus a Gaussian measure P on the σ-algebra $\mathscr{A}(T)$ is uniquely defined by a mean value function $A(t)$, $t \in T$, and a covariance function $B(s, t)$, $s, t \in T$.

The mean value $A(t)$, $t \in T$, can be arbitrary but the covariance function $B(s, t)$, $s, t \in T$, must be positive definite, i.e.,

$$\sum_{k, j=1}^{n} c_k c_j B(t_k, t_j) \geq 0$$

for any $t_1, \ldots, t_n \in T$ and real $c_1, \ldots, c_n$. For each function $A(t)$, $t \in T$, and positive $B(s, t)$, $s, t \in T$, there is a Gaussian random function with mean value A and covariance function B. Namely, the Gaussian distributions $P_{t_1, \ldots, t_n}$ with mean values $(A(t_1), \ldots, A(t_n))$ and covariance matrices $\{B(t_k, t_j)\}$ are consistent and give a Gaussian measure P on the space $X = \mathbb{R}^T$ of real functions $x = x(t)$ of $t \in T$, defined on the σ-algebra generated by cylinder sets.

A random variable ξ with values in the n-dimensional space $\mathbb{R}^n$ is *Gaussian* if its probability distributions are Gaussian. Clearly a random variable $\xi \in \mathbb{R}^n$ is Gaussian if and only if the real random variable $\xi(u) = (u, \xi)$, the scalar product of u and ξ in $\mathbb{R}^n$, is Gaussian for every $u \in \mathbb{R}^n$. The characteristic function $\varphi(u)$, $u \in \mathbb{R}^n$, for the random variable $\xi \in \mathbb{R}^n$ coincides on $u \in \mathbb{R}^n$ with the value of the characteristic function of the random variable $\xi(u) = (u, \xi)$ and has the form

$$\varphi(u) = E \exp\{i(u, \xi)\} = \exp\{i(u, A) - \tfrac{1}{2}(Bu, u)\}, \qquad u \in \mathbb{R}^n.$$

Thus it is also clear that the random variable $\xi \in \mathbb{R}^n$ is Gaussian if and only if the random function $\xi(u) = (u, \xi)$ of $u \in \mathbb{R}^n$ is Gaussian.

Naturally, the linear span of Gaussian random variables is Gaussian. If ξ is the L^2 limit of a sequence of Gaussian random variables $\xi^{(p)}$, $p = 1, 2, \ldots,$ then ξ is Gaussian; more precisely, if every collection of random variables $(\xi_1^{(p)}, \ldots, \xi_n^{(p)})$ is Gaussian, then the limit as $p \to \infty$, $(\xi_1, \ldots, \xi_n)$, is Gaussian also. This follows from the general expression for characteristic functions (5.1) if we note that as $p \to \infty$

$$E\xi_k^{(p)} \to E\xi \quad \text{and} \quad E\xi_k^{(p)}\xi_j^{(p)} \to E\xi_k \xi_j, \qquad j, k = 1, \ldots, n,$$

and the characteristic function of the limit variables $(\xi_1, \ldots, \xi_n)$ is the limit of the characteristic functions of the variables $(\xi_1^{(p)}, \ldots, \xi_n^{(p)})$, $p = 1, 2, \ldots$. Thus the L^2 closure of the linear span of Gaussian variables is Gaussian.

When we look at Gaussian variables as elements of the Hilbert space $L^2(\Omega, \mathscr{A}, P)$ we observe an important property: *the orthogonality of Gaussian random variables (with mean zero) is equivalent to their independence.*

This follows immediately from the general form of the characteristic function of a multi-dimensional Gaussian distribution: for arbitrary orthogonal collections $(\xi_1, \ldots, \xi_m)$ and $(\xi_{m+1}, \ldots, \xi_n)$, the characteristic function (5.1) of the variables $(\xi_1, \ldots, \xi_n)$ factors into the product of characteristic functions of $(\xi_1, \ldots, \xi_m)$ and $(\xi_{m+1}, \ldots, \xi_n)$.

A random element ξ in the Hilbert space U, i.e., a function $\xi = \xi(\omega)$ on the probability space $(\Omega, \mathscr{A}, P)$ with values in U, is a *random variable in U* if for every $u \in U$, the scalar product (u, ξ) is a real random variable—a measurable function on $(\Omega, \mathscr{A}, P)$; in other words, if the function $\xi = \xi(\omega)$, $\omega \in \Omega$, is weakly measurable.

A random variable ξ in the Hilbert space U is *Gaussian* if for every $u \in U$, the real variable $\xi(u) = (u, \xi)$ is Gaussian. This is equivalent to saying that the random function $\xi(u) = (u, \xi)$ of $u \in U$ is Gaussian, since not only the individual variables $\xi(u) = (u, \xi)$, but also all vector-valued variables $(\xi(u_1), \ldots, \xi(u_n))$ are Gaussian. In fact, for any vector $\lambda = (\lambda_1, \ldots, \lambda_n)$ in $\mathbb{R}^n$ the scalar product of λ with $(\xi(u_1), \ldots, \xi(u_n))$ becomes

$$\sum_{k=1}^{n} \lambda_k \xi(u_k) = \xi\left(\sum_{k=1}^{n} \lambda_k u_k\right) = \xi(u),$$

where $u = \sum_{k=1}^{n} \lambda_k u_k \in U$, and by our conditions the variable $\xi(u) = (u, \xi)$ is Gaussian.

Each element $x \in U$ of the Hilbert space U can be identified with the function on U given by $x(u) = (u, x)$—a continuous linear functional on U. Similarly the random variable $\xi \in U$ can be identified with a random continuous linear functional $\xi(u) = (u, \xi)$ on U. The mean value

$$A(u) = E(u, \xi), \qquad u \in U,$$

is a linear functional and the correlation function

$$B(u, v) = E[(u, \xi) - A(u)][(v, \xi) - A(v)], \qquad u, v \in U,$$

is a positive bilinear functional.

Integration of a Gaussian Function. We consider a Gaussian function $\xi(t)$, $t \in T$, on a measurable space T with measure $\mu(dt)$ as a function with values in the space $L^2(\Omega, \mathscr{A}, P)$ and assume it is (strongly) integrable. Recall that a piecewise-constant function $\xi(t)$ taking values ξ_k on measurable sets $\Delta_k \subseteq T$, where $\bigcup_k \Delta_k = T$,

$$\xi(t) = \xi_k, \qquad t \in \Delta_k; \qquad k = 1, 2, \ldots,$$

is integrable if the real-valued function $\|\xi(t)\|$, $t \in T$, is integrable; in this case

$$\int_T \xi(t)\mu(dt) = \sum_k \xi_k \mu(\Delta_k).$$

An arbitrary random function $\xi(t)$ is called *strongly integrable* if there is some sequence of integrable piecewise-constant functions $\xi_n(t)$ such that

$$\lim \int_T \|\xi(t) - \xi_n(t)\|\mu(dt) = 0;$$

in this case the (strong) limit

$$\int_T \xi(t)\mu(dt) = \lim \int_T \xi_n(t)\mu(dt)$$

does not depend on the choice of approximating sequence. It is clear that the approximating functions $\xi_n(t)$ always can be taken with values in the closed linear span of the variables $\xi(t)$, $t \in T$, (let us denote it by $H(T)$), since for the projections $\mathring{\xi}_n(t)$ of the variables $\xi_n(t)$ on $H(T)$ we have

$$\|\xi(t) - \mathring{\xi}_n(t)\| \le \|\xi(t) - \xi_n(t)\|.$$

Thus

$$\int_T \xi(t)\mu(dt) \in H(T), \tag{5.3}$$

and for Gaussian variables $\xi(t)$, $t \in T$, the set $H(T)$ is Gaussian.

We remark that the function

$$\xi(t) \in L^2(\Omega, \mathscr{A}, P), \qquad t \in T,$$

is integrable if and only if the real function $\xi(t) = \xi(\omega, t)$ is equivalent to some function of the pair of variables $(\omega, t) \in \Omega \times T$ measurable with respect to the product measure $P \times \mu$, and

$$\int_T \|\xi(t)\|\mu(dt) < \infty.$$

A suitable approximating sequence is obtained by taking real piecewise-constant functions $\xi_n(t) = \xi_n(\omega, t)$ taking a finite number of different values on disjoint sets $A_{nk} \times \Delta_{nk}$, where $A_{nk} \in \Omega$ and $\Delta_{nk} \in T$ are measurable sets (as we have seen, the collection of such sets form a semi-ring which generates a σ-algebra of measurable sets in the product $\Omega \times T$).

Then the integral $\int \xi(t)\mu(dt)$, regarded as a variable in the space $L^2(\Omega, \mathscr{A}, P)$ is

$$\int_T \xi(t)\mu(dt) = \int_T \xi(\omega, t)\mu(dt), \qquad \omega \in \Omega, \tag{5.4}$$

where the right side is the integral of the real-valued function $\xi(\omega, t)$, $t \in T$, for each fixed ω; this integral exists for almost all $\omega \in \Omega$, since

$$\int_\Omega \int_T |\xi(\omega, t)|\mu(dt)P(d\omega) \le \int_T \|\xi(t)\|\mu(dt).$$

We will consider the case when

$$\int_T \|\xi(t)\|^2\mu(dt) = \iint_{\Omega \times T} |\xi(\omega, t)|^2 P(d\omega)\mu(dt), \tag{5.5}$$

that is, when the function $\xi(\omega, t)$ of the variables (ω, t) belongs to the space $L^2(\Omega \times T, P \times \mu)$. Under condition (5.5), for almost all ω

$$\int_T |\xi(\omega, t)|^2\mu(dt) < \infty. \tag{5.6}$$

Let N be the set of ω for which (5.6) fails. We change $\xi(\omega, t)$ on the exceptional set $N \times T$ of measure 0 (by setting $\xi(\omega, t) = 0$, for instance). The resulting random function $\xi(t)$, $t \in T$, which is equivalent to the original one, satisfies (5.6) for all $\omega \in \Omega$ and its trajectory

$$\xi(\omega, \cdot) = \xi(\omega, t), \qquad t \in T,$$

considered as a function of $\omega \in \Omega$ is a random variable in the Hilbert space $U = L^2(T, \mu)$ of real functions $u = u(t)$, $t \in T$, which are square-integrable with respect to $\mu(dt)$. This random variable is Gaussian with mean value

$$A(u) = \int_T u(t)A(t)\mu(dt)$$

and correlation function

$$B(u, v) = \iint_{T \times T} u(s)v(t)B(s, t)\mu(ds)\mu(dt), \qquad u, v \in U,$$

where $A(t)$ and $B(s, t)$ are the mean value and correlation function for $\xi(t)$, $t \in T$.

2. Some Useful Propositions

(a) *Let H_n, $n = 1, 2, \ldots$, be a sequence of linear spaces of Gaussian variables, closed in mean square, let*

$$H = \bigcap_n H_n,$$

and let $\mathscr{A}$ be the σ-algebra generated by variables in the space H. Then

$$\bigcap_n \mathscr{A}_n = \mathscr{A}, \tag{5.7}$$

where $\mathscr{A}_n$, $n = 1, 2, \ldots$, are σ-algebras generated by the corresponding spaces H_n.

PROOF. Without loss of generality we can assume that $H_n = \bigcap_{k \le n} H_k$ and hence that $H_m \supseteq H_n$ for $m < n$. It is essential that the space

$$\Delta_n = H_{n-1} \ominus H_n$$

(the orthogonal complement in H_{n-1} of $H_n \subseteq H_{n-1}$) is composed of variables which are independent of H_n. Let $\mathscr{B}_n$ be the σ-algebra generated by the space Δ_n. We have

$$\mathscr{A}_n = \mathscr{A} \bigvee_{k \ge n} \mathscr{B}_k,$$

where $\mathscr{B}_k$, $k = 1, 2, \ldots$, is a sequence of independent σ-algebras. Let $\mathscr{B} = \bigvee_{k=1}^{\infty} \mathscr{B}_k$; $\mathscr{A}$ and $\mathscr{B}$ are independent σ-algebras. As we know, the σ-algebra $\mathscr{A} \vee \mathscr{B}$ is generated by events of the form $A \cdot B$, where $A \in \mathscr{A}$ and $B \in \mathscr{B}$, and the space $L^2(\mathscr{A} \vee \mathscr{B})$ is generated by corresponding indicator functions $1_A \cdot 1_B$; all possible products $\xi \cdot \eta$, $\xi \in L^2(\mathscr{A})$, $\eta \in L^2(\mathscr{B})$, form a complete system in the space $L^2(\mathscr{A} \vee \mathscr{B})$ and by independence of ξ and η

$$\|\xi\eta\|^2 = E|\xi\eta|^2 = E|\xi|^2 \cdot E|\eta|^2 = \|\xi\|^2 \cdot \|\eta\|^2.$$

In the decomposition

$$\xi\eta = (E\eta) \cdot \xi + \xi(\eta - E\eta)$$

the first term is in $L^2(\mathscr{A})$ and the second is orthogonal to it and therefore the variables

$$\xi \cdot (\eta - E\eta), \qquad \xi \in L^2(\mathscr{A}), \qquad \eta \in L^2(\mathscr{B}),$$

form a complete system in the orthogonal complement

$$L^2(\mathscr{A})^{\perp} = L^2(\mathscr{A} \vee \mathscr{B}) \ominus L^2(\mathscr{A}).$$

Moreover, the collection of products $\xi(\eta - E\eta)$ is a complete system in $L^2(\mathscr{A})^{\perp}$ whenever $\xi \in L^2(\mathscr{A})$ and $\eta \in L^2(\mathscr{B})$ are complete systems.
    We consider an arbitrary variable

$$\zeta \in \bigcap_n L^2(\mathscr{A}_n), \qquad \zeta \perp L^2(\mathscr{A}).$$

Since ζ is measurable with respect to $\mathcal{A}_n$ for any $n = 1, 2, \ldots$, it does not depend on $\mathcal{B}_1, \mathcal{B}_2, \ldots$, and thus is independent of the variables η. The product $\xi\bar{\zeta}, \xi \in L^2(\mathcal{A})$, also has this property and since our system of variables $\xi(\eta - E\eta)$ is complete in the orthogonal complement $L^2(\mathcal{A})^\perp$, we have

$$(\xi(\eta - E\eta), \zeta) = E(\eta - E\eta) \cdot E\xi\bar{\zeta} = 0,$$

that is, the variable ζ is orthogonal to all the variables $\xi(\eta - E\eta)$, hence $\zeta = 0$. From this,

$$L^2(\mathcal{A}) = \bigcap_n L^2(\mathcal{A}_n),$$

which is equivalent to (5.7). $\square$

(b) *Let us consider a sequence of random variables*

$$\xi_n = \xi_n(\omega), \qquad n = 1, 2, \ldots,$$

on the probability space $(\Omega, \mathcal{A}, P)$. We will say that the sequence $\xi_n, n = 1, 2, \ldots$, converges in probability to the variable $\xi = \xi(\omega)$ on the set $A \in \mathcal{A}$ if for every $\varepsilon > 0$,

$$\lim_{n \to \infty} P(\{|\xi_n - \xi| > \varepsilon\} \cap A) = 0. \tag{5.8}$$

We know that a sequence $\xi_n, n = 1, 2, \ldots$, converges in probability if and only if it is Cauchy, i.e., on the same set A, the sequence $\Delta_{nm} = \xi_n - \xi_m$, $n, m = 1, 2, \ldots$, converges in probability to 0.

The following is true:

If a sequence of Gaussian variables $\xi_n, n = 1, 2, \ldots$, converges in probability on some set $A \in \mathcal{A}$ of positive measure $(P(A) > 0)$, then it converges in L^2 (i.e., in mean square):

$$\lim_{n \to \infty} E[\xi_n - \xi]^2 = 0.$$

PROOF. Consider the Gaussian variables $\Delta_{nm} = \xi_n - \xi_m$. For every $\varepsilon > 0$

$$P\{|\Delta_{nm}| > \varepsilon\} = 2 \int_\varepsilon^\infty \frac{1}{\sqrt{2\pi}\,\sigma_{nm}} \exp\left\{-\frac{(x - a_{nm})^2}{2\sigma_{nm}^2}\right\} dx,$$

where $a_{nm} = E\Delta_{nm}$ and $\sigma_{nm}^2 = E(\Delta_{nm} - a_{nm})^2$. If the sequence ξ_n does not converge in mean square, this is equivalent to

$$\varlimsup_{n, m \to \infty} (a_{nm}^2 + \sigma_{nm}^2) > 0.$$

In that case, for sufficiently small $\varepsilon > 0$

$$\varlimsup_{n, m \to \infty} P\{|\Delta_{nm}| > \varepsilon\} \geq 1 - p/2,$$

where $p = P(A) > 0$. But then

$$\varlimsup_{n, m \to \infty} P(\{|\Delta_{nm}| > \varepsilon\} \cap A) \geq p/2,$$

which contradicts (5.8). Thus

$$\varlimsup_{n, m \to \infty} E\Delta_{nm}^2 = \lim_{n, m \to \infty} (a_{nm}^2 + \sigma_{nm}^2) = 0;$$

that is, the sequence ξ_n is L^2-Cauchy and consequently converges in mean square.

In particular, *if a sequence of Gaussian variables ξ_n converges on a set of positive measure (i.e., converges for all ω belonging to a set $A \in \mathcal{A}$ of positive measure), then it converges in mean square.* □

(c) Let $\xi_n, n = 1, 2, \ldots$, *be a sequence of independent Gaussian variables. The series $\sum_{n=1}^{\infty} \xi_n^2$ converges with positive probability if and only if the series $\sum_{n=1}^{\infty} E\xi_n^2$ converges.*

PROOF. Since

$$\sum_{n=1}^{\infty} E\xi_n^2 = E \sum_{n=1}^{\infty} \xi_n^2,$$

then convergence of the series $\sum_{n=1}^{\infty} E\xi_n^2$ implies that the random variable $\sum_{n=1}^{\infty} \xi_n^2(\omega)$ is finite for almost all $\omega \in \Omega$, that is, the series $\sum_{n=1}^{\infty} \xi_n^2$ converges with probability 1. Now let the series $\sum_{n=1}^{\infty} \xi_n^2$ converge with positive probability. Then with equal or greater probability, $\xi_n \to 0$ and thus $E\xi_n^2 \to 0$ as $n \to \infty$. We have

$$0 < E \exp\left\{-\sum_{n=1}^{\infty} \xi_n^2\right\} = \prod_{n=1}^{\infty} E \exp(-\xi_n^2)$$

$$\leq \prod_{n=1}^{\infty} E(1 - \xi_n^2 + \tfrac{1}{2}\xi_n^4) = \prod_{n=1}^{\infty} [1 - E\xi_n^2 + o(E\xi_n^2)],$$

which is possible only if $\sum_{n=1}^{\infty} E\xi_n^2 < \infty$. □

3. Gaussian Linear Functionals on Countably-Normed Hilbert Spaces

Let U be an arbitrary complete separable countably-normed Hilbert space with a system of scalar products

$$(u, v)_1, (u, v)_2, \ldots,$$

such that the norms $\|u\|_n = \sqrt{(u, u)_n}$ are monotonically nondecreasing and U is a linear topological space with neighborhood basis at the origin of the form

$$\{u: \|u\|_n < \varepsilon\}, \qquad n = 1, \ldots, p, \ \varepsilon > 0.$$

(As an example, take the space $U = C_0^\infty(T)$ of infinitely differentiable functions $u = u(t)$ with compact support Supp $u \subseteq T$, T a bounded domain in $\mathbb{R}^d$). The space U can be regarded as the intersection of Hilbert spaces U_n, $n = 1, 2, \ldots$, each of which is the completion of U with respect to the norm $\|u\|_n$ defined in (1.12). The space X of all continuous linear functionals $x = (u, x)$ on U consists of the union of spaces X_n, each of which is dual to the corresponding Hilbert space U_n,

$$U = \bigcap_{n=1}^{\infty} U_n, \qquad X = \bigcup_{n=1}^{\infty} X_n,$$

(here, $U_1 \supseteq U_2 \supseteq \cdots$ and $X_1 \subseteq X_2 \subseteq \cdots$).

Let $\mathscr{B}$ be the minimal σ-algebra containing all cylinder sets in the space X—all possible sets of the form

$$\{(u_1, x) \in \Gamma_1, \ldots, (u_n, x) \in \Gamma_n\},$$

where $u_1, \ldots, u_n \in U$ and $\Gamma_1, \ldots, \Gamma_n$ are Borel sets in $\mathbb{R}^1$.

Let $(\Omega, \mathscr{A}, P)$ be a probability space. For every $\omega \in \Omega$, let there be defined an element $\xi(\omega)$ in the space X conjugate to the countably-normed Hilbert space U, that is, a continuous linear functional $(u, \xi) = (u, \xi(\omega))$, $u \in U$. In order that for each fixed $u \in U$, the real-valued functions $(u, \xi) = (u, \xi(\omega))$ of $\omega \in \Omega$, be measurable (in other words, be random variables on the probability space $(\Omega, \mathscr{A}, P)$), it is necessary and sufficient that the function $\xi = \xi(\omega)$ with values in the measurable space $(X, \mathscr{B})$ be measurable; here $\mathscr{B}$ is the minimal σ-algebra containing all cylinder sets in the space X. We will call the measurable function $\xi = \xi(\omega)$ a random variable in the space X—the space of continuous linear functions $x = (u, x)$ on the countably-normed Hilbert space U; $\xi \in X$ is a *random continuous linear functional on U*: a random function $\xi = (u, \xi)$ with parameter $u \in U$, each of whose trajectories $\xi(\omega, \cdot) = (u, \xi(\omega))$, $u \in U$, is a continuous linear functional on U.

A random function $\xi = (u, \xi)$ with parameter $u \in U$ will be called a *random linear functional* if for every $u_1, u_2 \in U$ and real λ_1, λ_2,

$$\begin{cases} (\lambda_1 u_1 + \lambda_2 u_2, \xi) = \lambda_1(u_1, \xi) + \lambda_2(u_2, \xi), & \text{a.e.} \\ E[(u, \xi) - (v, \xi)]^2 \to 0, \end{cases} \tag{5.9}$$

as $u \to v$ in the countably-normed Hilbert space U, that is, $\|u - v\|_m \to 0$ for each $m = 1, 2, \ldots$.

The following is easy to show: a Gaussian random function $\xi = (u, \xi)$ with parameter $u \in U$ is a random linear functional if and only if its mean value $A(u)$, $u \in U$, is a continuous linear functional on the countably-normed Hilbert space U and its covariance function $B(u, v)$, $u, v \in U$, is a continuous bilinear functional on U (we call this a *covariance functional*).

The mean value of a random linear functional admits a representation of the form

$$A(u) = (u, A)_n, \qquad u \in U,$$

where A is a well-defined element in one of the Hilbert spaces U_n. The covariance functional has a representation of the form

$$B(u, v) = (Bu, v)_n, \qquad u, v \in U,$$

where B is a symmetric, positive continuous linear operator on some Hilbert space U_n; we call such an operator a *covariance operator*. The random linear functional $\xi = (u, \xi)$, $u \in U$, can be extended by continuity on any Hilbert space U_n on which both the mean value and covariance functional are continuous; this will be the case for sufficiently large n. For each element $v \in U_n$ there is a limit value (v, ξ),

$$E[(v, \xi) - (u, \xi)]^2 \to 0 \quad \text{as } u \to v, u \in U, \tag{5.10}$$

and the random function $\xi = (v, \xi)$ with parameter $v \in U_n$ defined in this way is a random linear functional on the Hilbert space U_n.

A random continuous linear functional $\xi = (u, \xi)$ is *Gaussian* if its values (u, ξ), $u \in U$, are Gaussian random variables (in other words, if the random function ξ with parameter $u \in U$ is Gaussian).

We consider a Gaussian continuous linear functional $\xi = (u, \xi)$, $u \in U$. From the definition, for each fixed $\omega \in \Omega$, we have both

$$(\lambda_1 u_1 + \lambda_2 u_2, \xi) = \lambda_1 (u_1, \xi) + \lambda_2 (u_2, \xi)$$

for any $u_1, u_2 \in U$ and real numbers λ_1, λ_2, and also

$$(u, \xi) - (v, \xi) \to 0 \quad \text{as } u \to v.$$

From §5.2 we see that a Gaussian function $\xi = (u, \xi)$ with parameter $u \in U$ which satisfies these conditions must also satisfy (5.9); that is, a Gaussian continuous linear functional is simultaneously a random linear functional which is continuous in mean square.

A natural question arises: when is there a Gaussian continuous linear functional equivalent to a given Gaussian linear functional $\xi = (u, \xi)$?

In answering this, we may as well assume that the mean value $A(u)$, $u \in U$, (a continuous linear functional on the space U) is equal to 0, since we can consider, in place of (u, ξ), $u \in U$, the Gaussian continuous linear functional $(u, \xi) - A(u)$, $u \in U$.

Let $\xi = (u, \xi)$ be a Gaussian continuous linear functional with mean value 0 on the countably-normed Hilbert space U:

$$\xi(\omega) = (u, \xi(\omega)), \qquad u \in U,$$

(for each $\omega \in \Omega$, $\xi(\omega)$ belongs to X, the space of continuous linear functionals $x = (u, x)$, $u \in U$, on the space U). Recall that $X = \bigcup_{n=1}^{\infty} X_n$, where X_n is the dual of the Hilbert space U_n which is obtained by completing U with respect to the corresponding norm $\|u\|_n$. Then with positive probability the random variable $\xi \in X$ is in some Hilbert space X_n, $\xi(\omega) \in X_n$ for $\omega \in A$, $P(A) > 0$, and

$$(u, \xi(\omega)) = (u, \xi^{(n)}(\omega))_n,$$

where $\xi^{(n)}(\omega)$ is an element thus determined in the Hilbert space U_n. Without loss of generality we can assume that the covariance operator for our random linear functional $\xi = (u, \xi)$ is defined on this space U_n as well. Now consider the Gaussian variables (v, ξ), $v \in U_n$, defined by the limit (5.10). For any orthonormal system $\{v_1, v_2, \ldots\}$ in the Hilbert space U_n the corresponding Gaussian variables $\xi_k = (v_k, \xi)$, $k = 1, 2, \ldots$, converge to zero as $k \to \infty$ on the set $\omega \in A$:

$$\xi_k(\omega) = (v_k, \xi^{(n)}(\omega))_n \to 0, \qquad k \to \infty.$$

Consequently they also converge in mean square:

$$E\xi_k^2 = (Bv_k, v_k) \to 0.$$

This implies that the covariance operator B is completely continuous and moreover we can choose the complete orthonormal system $v_1, v_2, \ldots$ to be the eigenvectors of this continuous operator corresponding to its eigenvalues $\sigma_1^2, \sigma_2^2, \ldots$. Then the corresponding variables $\xi_k = (v_k, \xi)$, $k = 1, 2, \ldots$, are orthogonal:

$$E\xi_k\xi_j = (Bv_k, v_j) = \begin{cases} \sigma_k^2, & \text{if } j = k, \\ 0, & \text{if } j \neq k. \end{cases}$$

Thus for all $\omega \in A$

$$\sum_{k=1}^{\infty} \xi_k^2 = \sum_{k=1}^{\infty} (v_k, \xi^{(n)}(\omega))^2 = \|\xi^{(n)}\|_n^2.$$

The orthogonal Gaussian variables ξ_k, $k = 1, 2, \ldots$, are independent and from the convergence with positive probability $P(A)$ of the series $\sum_{k=1}^{\infty} \xi_k^2$ we have convergence of the series $\sum_{k=1}^{\infty} E\xi_k^2$ (cf. §5.2), and thus

$$\sum_{k=1}^{\infty} E\xi_k^2 = \sum_{k=1}^{\infty} (Bv_k, v_k) = \sum_{k=1}^{\infty} \sigma_k^2 < \infty.$$

This implies that the covariance operator B is *nuclear*.

Now let $\xi = (u, \xi)$ be a random linear functional having mean value zero and nuclear covariance operator B in some Hilbert space U_n. Let $v_1, v_2, \ldots$ be a complete orthonormal system in the Hilbert space U_n. As before, let $\xi_k = (v_k, \xi)$, $k = 1, 2, \ldots$. We have

$$E \sum_{k=1}^{\infty} \xi_k^2 = \sum_{k=1}^{\infty} E\xi_k^2 = \sum_{k=1}^{\infty} (Bv_k, v_k)$$

and thus for almost all $\omega \in \Omega$

$$\sum_{k=1}^{\infty} \xi_k^2(\omega) < \infty.$$

We define a random variable $\xi^{(n)} = \xi^{(n)}(\omega)$ in the Hilbert space U_n in such a way that for almost all $\omega \in \Omega$

$$\xi^{(n)}(\omega) = \sum_{k=1}^{\infty} \xi_k(\omega) v_k;$$

the formula

$$\xi(\omega) = (u, \xi^{(n)}(\omega))_n, \qquad u \in U, \tag{5.11}$$

defines for each $\omega \in \Omega$ a continuous linear functional $\xi = (u, \xi^{(n)})_n$, $u \in U_n$, on the countably-normed Hilbert space U such that $\xi \in X_n$. This continuous random linear functional $\xi \in X_n$ is equivalent to the original random functional $\xi = (u, \xi)$, $u \in U$, since for each fixed $u \in U$, with probability 1

$$(u, \xi^{(n)})_n = \sum_{k=1}^{\infty} \xi(v_k)(u, v_k) = \lim_{n \to \infty} \xi\left(\sum_{k=1}^{n} (u, v_k) v_k \right) = (u, \xi);$$

this follows from condition (5.9)—linearity and continuity in mean square.

Thus, *for every Gaussian random linear functional $\xi = (u, \xi)$, $u \in U$, there exists an equivalent Gaussian random continuous linear functional $\xi \in X$ if and only if the covariance operator B is nuclear on some Hilbert space U_n.* In this case, the Gaussian continuous linear functional $\xi \in X$ belongs to the space $X_n \subseteq X$ (and dual to U_n) with probability 1, and it can be described by formula (5.11).

4. Polynomials of Gaussian Variables and Their Conditional Expectations

Let $\xi(t)$, $t \in T$, be an arbitrary set of Gaussian variables on a probability space $(\Omega, \mathscr{A}, P)$. For a set $S \subseteq T$ we let

$$H(S) = L^2(\Omega, \mathscr{A}(S), P), \tag{5.12}$$

where $\mathscr{A}(S)$ is the σ-algebra generated by the variables $\xi(t)$, $t \in S$, and we let $H^n(S)$ denote the closed linear span of all polynomials in the variables $\xi(t)$, $t \in S$, of degree $\leq n$; each of these polynomials is a linear combination of variables of the form

$$\xi(t_1) \cdots \xi(t_k), \qquad k \leq n \text{ and } t_1, \ldots, t_k \in S. \tag{5.13}$$

Variables $\eta \in H^n(T)$ have the following property:

$$E(\eta \mid \mathscr{A}(S)) \in H^n(S). \tag{5.14}$$

PROOF. We consider the case $n = 1$. A variable $\eta \in H^1(T)$ belongs to the closed linear span of the Gaussian variables $\xi(t)$, $t \in T$. We let $\hat{\eta}$ denote its projection on $H^1(S) \subseteq H^1(T)$. The Gaussian variable $\eta - \hat{\eta}$ is orthogonal to $H^1(S)$, does not depend on $H^1(S)$, and consequently is orthogonal to the whole space

$H(S)$, so that $\hat{\eta} \in H^1(S)$ is the projection of the variable η on the whole space $H(S)$:

$$\hat{\eta} = E(\eta \mid \mathscr{A}(S)).$$

Turning to the case $n > 1$, we first note that the conditional expectation $E(\eta \mid \mathscr{A}(S))$, like any other random variable measurable with respect to the σ-algebra $\mathscr{A}(S)$, is also measurable with respect to "its own" σ-algebra $\mathscr{A}(S')$, where $S' \subseteq S$ and is countable, and

$$E(\eta \mid \mathscr{A}(S)) = E(\eta \mid \mathscr{A}(S')) = \lim_n E(\eta \mid \mathscr{A}(S_n))$$

for some sequence of finite sets $S_n \subseteq S'$. It is clear that to show (5.14) one can take the set S to be finite; in this case the space $H^n(S)$ is finite dimensional and consists of linear combinations of variables (5.13). It is also clear that it is sufficient to consider variables of the form $\eta = \xi(t_1) \cdots \xi(t_n)$. We prove (5.14) by induction on m. Assume it is true for $n < m$. Let $\hat{\xi}(t_k)$ be the projection of each of the factors $\xi(t_k)$ on the subspace $H^1(S)$ and let

$$\zeta = (\xi(t_1) - \hat{\xi}(t_1)) \cdots (\xi(t_m) - \hat{\xi}(t_m)).$$

The variable ζ, together with the differences $\xi(t_k) - \hat{\xi}(t_k)$, $k = 1, \ldots, m$, is independent of the collection of variables $\xi(t)$, $t \in S$, and

$$E(\zeta \mid \mathscr{A}(S)) = E\zeta.$$

The difference $\eta - \zeta$ is a linear combination of products $\eta_1 \cdot \eta_2$ of variables of the form

$$\eta_1 = \xi(t_{k_1}) \cdots \xi(t_{k_n}), \qquad \eta_2 = \hat{\xi}(t_{k_{n+1}}) \cdots \hat{\xi}(t_{k_m}), \qquad n < m,$$

for which

$$E(\eta_1 \cdot \eta_2 \mid \mathscr{A}(S)) = E(\eta_1 \mid \mathscr{A}(S)) \cdot \eta_2,$$

and by our assumption that (5.14) is true for $n < m$,

$$\hat{\eta}_1 = E(\eta_1 \mid \mathscr{A}(S)) \in H^n(S)$$

is a polynomial of degree n in the variables $\xi(t)$, $t \in S$, and $\eta_2 \in H^{m-n}(S)$ is a polynomial of degree $m - n$ in these variables. Thus the space $H^m(S)$ contains not only the product $\hat{\eta}_1 \cdot \eta_2$ but also the variable

$$E(\eta \mid \mathscr{A}(S)) = E\zeta + E(\eta - \zeta \mid \mathscr{A}(S)),$$

which is what we wished to show. $\square$

The *Hermite polynomials* $\varphi(x)$ of one real argument $x \in \mathbb{R}^1$ result from extracting an orthogonal basis in the space $L^2(\mathbb{R}^1, \mathscr{B}, P)$ from the vectors $1, x, x^2, \ldots$. Here $\mathscr{B}$ is the Borel σ-algebra and P is a Gaussian measure having density

$$\frac{P(dx)}{dx} = \frac{1}{\sqrt{2\pi}} \exp(-x^2/2), \qquad x \in \mathbb{R}^1. \tag{5.15}$$

For an arbitrary Gaussian measure P on the k-dimensional space $\mathbb{R}^k$, a polynomial $\varphi(x_1, \ldots, x_k)$ of degree p in the variables $(x_1, \ldots, x_k) \in \mathbb{R}^k$ will be called *Hermite* if it is orthogonal to all polynomials of degree $q < p$ in the space $L^2(\mathbb{R}^k, \mathscr{B}^k, P)$.

Now consider an arbitrary collection of Gaussian variables $\xi(t)$, $t \in T$. Let $S \subseteq T$ be a finite set, say $S = \{t_1, \ldots, t_k\}$. We let $H_p(S)$ denote the collection of all Hermite polynomials of degree p

$$\eta = \varphi(\xi(t_1), \ldots, \xi(t_k))$$

corresponding to the joint distribution P of the variables $\xi(t_1), \ldots, \xi(t_k)$. Since polynomials form a complete set in the space $L^2(\mathbb{R}^k, \mathscr{B}^k, P)$, we can represent the space $H(S)$ (see (5.12)) as an orthogonal sum of subspaces $H_p(S)$, $p = 0$, $1, \ldots$:

$$H(S) = \sum_{p=0}^{\infty} \oplus H_p(S). \tag{5.16}$$

Let $S_1, S_2 \subseteq T$ be arbitrary (finite) sets and $H_p(S_1), H_q(S_2)$ be corresponding subspaces of Hermite polynomials of degree p and q. Recall that $H_p(S_1)$ and $H_q(S_2)$ are defined by orthogonalization processes on the corresponding spaces $H(S_1)$ and $H(S_2)$. At first glance, it might seem that there is no apparent relationship between $H_p(S_1)$ and $H_q(S_2)$ for various S_1 and S_2. However,

$$H_p(S_1) \perp H_q(S_2), \qquad p \neq q. \tag{5.17}$$

We will show this. Assume $p < q$. Consider an arbitrary variable $\eta \in H_p(S_1)$ and its conditional expectation $\hat{\eta} = E(\eta \,|\, \mathscr{A}(S_2))$. As we have seen, the variable $\hat{\eta}$ belongs to the subspace $H^p(S_2) = \sum_{r=0}^{p} \oplus H_r(S_2)$. But $\hat{\eta}$ is the projection of the variable η on the whole space $H(S_2)$, so that the difference $\eta - \hat{\eta}$ is orthogonal to $H(S_2)$ and in particular, $\eta - \hat{\eta} \perp H_q(S_2)$. At the same time, $\hat{\eta} \perp H_q(S_2)$, since $\hat{\eta}$ belongs to the subspace $H^p(S_2)$ which is orthogonal to $H_q(S_2)$ for $p < q$ by construction. Thus we have established that

$$\eta = [(\eta - \hat{\eta}) + \hat{\eta}] \perp H_q(S_2). \qquad \square$$

General Orthogonal Expansion. For arbitrary $S \subseteq T$, we define the space $H_p(S)$ as the closed linear span of all Hermite polynomials of degree p in the variables $\xi(t)$, $t \in S$. It is clear that relations (5.16) and (5.17) extend to arbitrary $S \subseteq T$. In (5.16), the orthogonal sum of the first n subspaces H_p, $p \leq n$, gives us

$$H^n(S) = \sum_{p=0}^{n} \oplus H_p(S). \tag{5.18}$$

The following proposition holds:

For all $S \subseteq T$, each variable $\eta \in H(S)$ has a projection on the subspace $H_p(T)$ which coincides with the projections $\eta_p(S)$ of η on the subspaces $H_p(S)$.

PROOF. In the expansion

$$\eta = \sum_{q=0}^{\infty} \oplus \eta_q(S)$$

each of the variables $\eta_q(S)$, $q \neq p$, is orthogonal to the subspace $H_p(T)$ by (5.17); consequently $\eta_p(S) \in H_p(T)$ is the projection of the variable η on $H_p(T)$.

$\square$

From the expansion (5.18) applied to $S = T$ and the general relationship (5.17) it immediately follows that the projection of the space $H_p(T)$ on the subspace $H(S)$ is $H_p(S)$. This sharpens the property of conditional expectations given in (5.14), namely,

$$E(\eta \mid \mathscr{A}(S)) \in H_p(S) \qquad (5.19)$$

for any variable $\eta \in H_p(T)$.

We know the set

$$H(T) = H_1(T);$$

that is, the closed linear span of Gaussian variables $\xi(t)$, $t \in S$, having mean value zero is Gaussian. It is not hard to check that the closed linear span of polynomials in variables $\eta \in H(T)$, of degree $\leq n$, coincides with the space $H^n(T)$. In view of this, one can pass from an arbitrary Gaussian family $\xi(t)$, $t \in T$, to a system of orthogonal Gaussian variables by taking an orthogonal basis of the space $H(T)$.

Starting with a given orthonormal system of Gaussian variables $\xi(t)$, $t \in T$, we can specify the structure of the subspace $H_p(T)$ generated by the Hermite polynomials of degree p. Each of the variables $\xi(t)$ has the probability density (5.15) and the distribution P of the variables $(\xi(t_1), \ldots, \xi(t_k))$ in $\mathbb{R}^k$ is the product distribution of the separate components $\xi(t_j)$, $j = 1, \ldots, k$, in $\mathbb{R}^1$. Therefore, if $\varphi_p(x)$ denotes the Hermite polynomial of degree p of a real variable $x \in \mathbb{R}^1$ and

$$\varphi_p(x), \qquad p = 0, 1, \ldots,$$

is a complete orthonormal set of these polynomials, then all possible products

$$\varphi_{p_1}(x_1) \cdots \varphi_{p_k}(x_k), \qquad p_i = 0, 1, \ldots,$$

are Hermite polynomials of degree $p = p_1 + \cdots + p_k$ in variables $(x_1, \ldots, x_k)$ $\in \mathbb{R}^k$ and form a complete orthonormal basis of the space $L^2(\mathbb{R}^k, \mathscr{B}^k, P)$. From this we get that for any finite subset $S = \{t_1, \ldots, t_k\} \subseteq T$, the variables

$$\eta = \varphi_{p_1}(\xi(t_1)) \cdots \varphi_{p_k}(\xi(t_k)), \qquad p_1 + \cdots + p_k = p, \qquad (5.20)$$

form an orthonormal basis in the subspace $H_p(S)$. It is evident that this result holds for any subset $S \subseteq T$, since each variable $\eta \in H_p(S)$ will be measurable with respect to a σ-algebra generated by a countable number of variables $\xi(t_1)$, $\xi(t_2)$, $\ldots$, and can be approximated by Hermite polynomials of the form (5.20).

5. Hermite Polynomials and Multiple Stochastic Integrals

Let $\eta(\Delta)$, $\Delta \subseteq T$, be a Gaussian stochastic orthogonal measure on a domain $T \subseteq \mathbb{R}^d$ with

$$E|\eta(dt)|^2 = b(t) \, dt;$$

then

$$(u, \xi) = \int_T u(t)\eta(dt), \qquad u \in U,$$

is a Gaussian linear functional on the Hilbert space $U = L^2(T, \mathcal{B}, \mu)$, where $\mathcal{B}$ is a σ-algebra of measurable sets $\Delta \subseteq T$ and $\mu(dt) = b(t) \, dt$.

We let $H_p(U)$ denote the closed linear span of Hermite polynomials of degree p in the variables $\xi(u) = (u, \xi), u \in U$. The space $H_p(U)$ clearly coincides with the closure of the Hermite polynomials in $\eta(\Delta)$, $\Delta \subseteq T$, of degree p; however note that η is parametrized by $\Delta \subseteq T$ while ξ is parameterized by L^2 functions on T.

Using the additivity of the variables $\eta(\Delta)$, $\Delta \subseteq T$, each pth degree polynomial in these variables can be represented in the form

$$\sum_{k_1, \ldots, k_p} u(t_1, \ldots, t_p)\eta(\Delta_{k_1}) \cdots \eta(\Delta_{k_p}), \qquad (5.21)$$

where Δ_k, $k = 1, 2, \ldots$, are disjoint measurable sets forming a sufficiently fine partition of the region $T = \bigcup_k \Delta_k$ and $u(t_1, \ldots, t_p)$ is a piecewise-constant function of $(t_1, \ldots, t_p) \in T^p$, having a constant value on each set

$$\Delta_{k_1} \times \cdots \times \Delta_{k_p} \subseteq T^p, \qquad k_i = 1, 2, \ldots,$$

and differing from zero only on a finite number of these sets.

We note that when each product $\eta(\Delta_{k_1}) \cdots \eta(\Delta_{k_p})$ in (5.21) contains only distinct $\Delta_{k_1}, \ldots, \Delta_{k_p}$, and under the additional condition

$$u(t_1, \ldots, t_p) = 0 \quad \text{if} \quad (t_1, \ldots, t_p) \in \bigcup_{i \neq j} \{t_i = t_j\}, \qquad (5.22)$$

the variables defined in (5.21) are orthogonal to all polynomials of degree $q < p$. In fact, each polynomial of degree q is representable as a linear combination of variables of the form $\eta(\Delta'_1) \cdots \eta(\Delta'_q)$, where each set Δ'_j is contained in some Δ_k in the representation (5.21); but in this representation each product $\eta(\Delta_{k_1}) \cdots \eta(\Delta_{k_p})$ contains at least one component $\eta(\Delta_{k_i})$ for which Δ_{k_i} is disjoint from every $\Delta'_j, j = 1, \ldots, q$. Consequently

$$n(\Delta'_1) \cdots \eta(\Delta'_q) \perp \eta(\Delta_{k_1}) \cdots \eta(\Delta_{k_p}).$$

Recall here that the values of a Gaussian orthogonal measure have mean 0 and are independent for disjoint sets.

Under condition (5.22), the Hermite polynomials (5.21) can be represented in the form of *multiple stochastic integrals*:

$$
\begin{aligned}
I_p(u) &= \underset{T \times \cdots \times T}{\int \cdots \int} u(t_1, \ldots, t_p) \eta(dt_1) \cdots \eta(dt_p) \\
&= \sum_{k_1, \ldots, k_p} u(t_1, \ldots, t_p) \eta(\Delta_{k_1}) \cdots \eta(\Delta_{k_p}).
\end{aligned} \tag{5.23}
$$

The integral $I_p(u)$ defined by means of (5.23) takes the same value for all functions $u(t_1, \ldots, t_p)$ which differ from each other through a rearrangement of the variables $t_1, \ldots, t_p$ and

$$
I_p(u) = I_p(\tilde{u}),
$$

where

$$
\tilde{u}(t_1, \ldots, t_p) = \frac{1}{p!} \sum_{i_1, \ldots, i_p} u(t_{i_1}, \ldots, t_{i_p})
$$

is a symmetrization of the function $u(t_1, \ldots, t_p)$. It is easy to check that

$$
\begin{aligned}
\|I_p(\tilde{u})\|^2 &= E|I_p(\tilde{u})|^2 \\
&= p! \underset{T \times \cdots \times T}{\int \cdots \int} |\tilde{u}(t_1, \ldots, t_p)|^2 \mu(dt_1) \cdots \mu(dt_p) \\
&= p! \|\tilde{u}\|^2.
\end{aligned} \tag{5.24}
$$

The isometric (to within a constant factor) correspondence

$$
\tilde{u} \leftrightarrow I_p(\tilde{u})
$$

between piecewise-constant symmetric functions $\tilde{u}$ as elements of the space

$$
L^2 = L^2(T^p, \mathscr{B}^p, \mu^p)
$$

and stochastic integrals $I_p(\tilde{u})$ can be extended to the closure of such functions $\tilde{u}$, recalling that they must satisfy condition (5.22). Clearly, the closure coincides with the subspace of all symmetric functions $\tilde{u} \in L^2$, since the set $\bigcup_{i \neq j} \{t_i = t_j\}$ of points $(t_1, \ldots, t_p) \in T^p$ for which at least two coordinates coincide has measure 0.

For an arbitrary function $u \in L^2$, we define a *p-times iterated stochastic integral* $I_p(u)$ by taking the limit:

$$
\begin{aligned}
I_p(u) &= \underset{T \times \cdots \times T}{\int \cdots \int} u(t_1, \ldots, t_p) \eta(dt_1) \cdots \eta(dt_p) \\
&= I_p(\tilde{u}) = \lim_{n \to \infty} I_p(\tilde{u}_n),
\end{aligned} \tag{5.25}
$$

where $\tilde{u}_n, n = 1, 2, \ldots,$ is a sequence of piecewise-constant functions satisfying (5.22) and converging to $\tilde{u} \in L^2$.

As we have essentially shown, for each function $u \in L^2$ the integral $I_p(u)$ belongs to the space $H_p(U)$, the closure of Hermite polynomials of degree p in the variables $\eta(\Delta)$, $\Delta \subseteq T$. It turns out that $H_p(U)$ *coincides with the collection of all (p-times) iterated stochastic integrals*:

$$H_p(U) = \{I_p(u), u \in L^2\}. \tag{5.26}$$

For $p = 1$ this is obvious. Assume it is true for all indices $p \le n$. That is, each polynomial of degree $\le n$ is a limit in mean square of some linear combination of polynomials $I_p(u)$ of the form (5.23) for $p \le n$. Then each polynomial of degree $\le n + 1$ in the variables $\eta(\Delta)$, $\Delta \subseteq T$, is a limit of linear combinations of products of the form $I_p(u) \cdot \eta(\Delta)$, or more generally, of products $I_p(u) \cdot I_1(v)$. For a suitable, sufficiently fine partition $T = \bigcup_k \Delta_k$, each variable $I_p(u) \cdot I_1(v)$ has an asymptotic representation:

$$I_p(u) \cdot I_1(v)$$

$$= \sum_{k_1, \ldots, k_p} u(t_1, \ldots, t_p)\eta(\Delta_{k_1}) \cdots \eta(\Delta_{k_p}) \cdot \sum_k v(t_k)\eta(\Delta_k)$$

$$= \left[\sum_{k_{p+1} \neq k_1, \ldots, k_p} u(t_1, \ldots, t_p)v(t_{p+1})\eta(\Delta_{k_1}) \cdots \eta(\Delta_{k_p})\eta(\Delta_{k_{p+1}}) \right]$$

$$\oplus \left[\sum_{q=1}^{p} \sum_{k_{p+1} = k_q} u(t_1, \ldots, t_p)v(t_q)\eta(\Delta_{k_1}) \cdots \eta(\Delta_{k_q-1})\mu(\Delta_{k_q})\eta(\Delta_{k_q+1}) \cdots \eta(\Delta_p) \right]$$

$$\oplus o(1),$$

in which the first term is a $p + 1$-times iterated integral $I_{p+1}(u \cdot v)$ of the product $u(t_1, \ldots, t_p) \cdot v(t_{p+1})$, regarded as a function of the variables $(t_1, \ldots, t_{p+1})$, the second term is a polynomial of degree $p - 1$ in the variables $\eta(\Delta)$, $\Delta \subseteq T$, and the remaining term

$$o(1) = \sum_{q=1}^{p} \left[\sum_{k_{p+1} = k_q} u(t_1, \ldots, t_p)v(t_q)\eta(\Delta_{k_1}) \cdots \eta(\Delta_{k_q-1}) \right.$$

$$\left. \times [\eta(\Delta_{k_q})^2 - \mu(\Delta_{k_q})]\eta(\Delta_{k_q+1}) \cdots \eta(\Delta_{k_p}) \right]$$

is bounded by

$$\|o(1)\| \le C \sup_k \mu(\Delta_k)^{1/2} \sum_{q=1}^{p} (p!)^{1/2} \|u(t_1, \ldots, t_p) \cdot v(t_q)\|,$$

where the norm on the right-hand side is of $u(t_1, \ldots, t_p) \cdot v(t_q)$ regarded as a function of the variables $(t_1, \ldots, t_p)$ and an element of the space $L^2(T^p, \mathscr{B}^p, \mu^p)$. As an L^2 bound,

$$\|o(1)\| = (E|o(1)|^2)^{1/2},$$

this is easily derived by noting that corresponding to disjoint $\Delta_{k_1}, \ldots, \Delta_{k_p}$, the factors in the expression $o(1)$ are independent and for each $\Delta = \Delta_1$, $\Delta_2, \ldots$

$$E\eta(\Delta)^2 = \mu(\Delta), \qquad E[\eta(\Delta)^2 - \mu(\Delta)]^2 = 2\mu(\Delta)^2.$$

We see that by choosing successively finer partitions of the region T into disjoint sets Δ_k, $k = 1, 2, \ldots$, with measure $\mu(\Delta_k)$ for which

$$\sup_k \mu(\Delta_k) \to 0,$$

any variable of the form $I_p(u) \cdot I_1(v)$ can be arbitrarily closely approximated by an orthogonal sum of corresponding integrals $I_{p+1}(u \cdot v)$ and polynomials of degree $\leq p$, and thus the $p + 1$-times iterated integrals are the orthogonal complement of the space $H^p(U)$—the closure of all polynomials of degree $\leq p$— in the space $H^{p+1}(U)$. Note that $H^{p+1} = H_{p+1} \oplus H^p$ and we have just shown that $H^{p+1} = I_{p+1} \oplus H^p$. This establishes that $H_p(U)$ is the set of all p-times iterated stochastic integrals $I_p(u)$ for $p \leq n + 1$ under the assumption that this is true for $p \leq n$.   □

CHAPTER 2

Markov Random Fields

§1. Basic Definitions and Useful Propositions

1. Splitting σ-algebras

Let $\mathscr{A}_1$, $\mathscr{B}$, $\mathscr{A}_2$ be σ-algebras of events having the following relationship: if the outcomes of all events in $\mathscr{B}$ are known, events $A_2 \in \mathscr{A}_2$ are independent of events $A_1 \in \mathscr{A}_1$. More precisely, the σ-algebras $\mathscr{A}_1$ and $\mathscr{A}_2$ are conditionally independent with respect to $\mathscr{B}$; this gives the equation for conditional probabilities:

$$P(A_1 \cdot A_2 | \mathscr{B}) = P(A_1 | \mathscr{B}) \cdot P(A_2 | \mathscr{B}) \tag{1.1}$$

for any $A_1 \in \mathscr{A}_1$, $A_2 \in \mathscr{A}_2$. We say that the σ-algebra $\mathscr{B}$ *splits* $\mathscr{A}_1$ and $\mathscr{A}_2$ (or *is splitting*) if (1.1) holds for $\mathscr{A}_1$, $\mathscr{B}$, $\mathscr{A}_2$.

Throughout the following discussion, in considering random variables and their conditional expectations (and in particular, conditional probabilities), we will not distinguish between variables which coincide with probability 1; for the most part we will be considering variables in the spaces

$$L^p(\mathscr{A}) = L^p(\Omega, \mathscr{A}, P), \qquad p = 1, 2.$$

Under condition (1.1), for arbitrary variables $\xi_1 \in L^2(\mathscr{A}_1)$, $\xi_2 \in L^2(\mathscr{A}_2)$ the following is true:

$$E(\xi_1 \xi_2 | \mathscr{B}) = E(\xi_1 | \mathscr{B}) \cdot E(\xi_2 | \mathscr{B}). \tag{1.2}$$

This is because (1.2) is clearly true for indicator functions $\xi_1 = 1_{A_1}$ and $\xi_2 = 1_{A_2}$ for all possible $A_1 \in \mathscr{A}_1$, $A_2 \in \mathscr{A}_2$, and extends to their closed linear spans $L^2(\mathscr{A}_1)$, $L^2(\mathscr{A}_2)$ since the conditional expectation $E(\cdot|\mathscr{B})$ is a linear operator which is continuous for $\xi_1, \xi_2 \in L^2$ and $\xi_1 \cdot \xi_2 \in L^1$.

It is clear that properties (1.1) and (1.2) are equivalent. Moreover we can say that the σ-algebra $\mathscr{B}$ splits $\mathscr{A}_1$ and $\mathscr{A}_2$ if and only if (1.2) holds for every complete system of variables $\xi_1 \in L^2(\mathscr{A}_1)$ and $\xi_2 \in L^2(\mathscr{A}_2)$.

From this criterion we derive some simple propositions.

If the σ-algebra $\mathscr{B}$ splits $\mathscr{A}_1$ and $\mathscr{A}_2$, then it also splits the σ-algebras

$$\tilde{\mathscr{A}}_1 = \mathscr{A}_1 \vee \mathscr{B} \quad \text{and} \quad \tilde{\mathscr{A}}_2 = \mathscr{A}_2 \vee \mathscr{B}. \tag{1.3}$$

In fact, since the σ-algebra $\tilde{\mathscr{A}}_1 = \mathscr{A}_1 \vee \mathscr{B}$ is generated by the semi-ring of events of the form $A_1 \cdot B$, $A_1 \in \mathscr{A}_1$ and $B \in \mathscr{B}$, a complete system of variables in $L^2(\tilde{\mathscr{A}}_1)$ is given by $\tilde{\xi}_1 = \xi_1 \cdot \eta_1$, where $\xi_1 = 1_{A_1}$ and $\eta_1 = 1_{B_1}$, $B_1 \in \mathscr{B}$. Taking an analogous system $\tilde{\xi}_2 = \xi_2 \cdot \eta_2$ in $L^2(\tilde{\mathscr{A}}_2)$, we have

$$\begin{aligned} E(\tilde{\xi}_1 \tilde{\xi}_2|\mathscr{B}) &= \eta_1 \eta_2 \cdot E(\xi_1 \xi_2|\mathscr{B}) = \eta_1 \eta_2 \cdot E(\xi_1|\mathscr{B}) \cdot E(\xi_2|\mathscr{B}) \\ &= E(\tilde{\xi}_1|\mathscr{B}) \cdot E(\tilde{\xi}_2|\mathscr{B}). \qquad \square \end{aligned}$$

The following holds:

$\mathscr{B}$ splits the algebras $\mathscr{A}_1$, $\mathscr{A}_2$ if and only if the sequence $\mathscr{A}_1, \mathscr{B}, \mathscr{A}_2$ is Markov, i.e.,

$$P(A|\mathscr{A}_1 \vee \mathscr{B}) = P(A|\mathscr{B}), \qquad A \in \mathscr{A}_2, \tag{1.4}$$

In effect, the σ-algebra $\mathscr{B}$ splits $\mathscr{A}_1$ and $\mathscr{A}_2$ if and only if it splits $\mathscr{A}_1 \vee \mathscr{B}$ and $\mathscr{A}_2$, and without loss of generality we can assume that $\mathscr{B} \subseteq \mathscr{A}_1$. Accordingly, $\mathscr{A}_1 \vee \mathscr{B} = \mathscr{A}_1$, and (1.4) is equivalent to

$$E(\xi|\mathscr{A}_1) = E(\xi|\mathscr{B}) \tag{1.5}$$

for an $L^2(\mathscr{A}_2)$-complete set of random variables ξ.

When (1.5) holds, then for $\xi_2 = \xi$ and $\xi_1 \in L^2(\mathscr{A}_1)$,

$$\begin{aligned} E(\xi_1 \xi_2|\mathscr{B}) &= E[E(\xi_1 \xi_2|\mathscr{A}_1)|\mathscr{B}] = E[\xi_1 \cdot E(\xi_2|\mathscr{B})|\mathscr{B}] \\ &= E(\xi_2|\mathscr{B}) \cdot E(\xi_1|\mathscr{B}). \end{aligned}$$

On the other hand, from (1.2) we have

$$E(\xi_1 \cdot [\xi_2 - E(\xi_2|\mathscr{B})]|\mathscr{B}) = 0;$$

hence

$$E(\xi_1 \cdot [\xi_2 - E(\xi_2|\mathscr{B})]) = 0$$

for any $\xi_1 \in L^2(\mathscr{A}_1)$. It is evident that for $\mathscr{B} \subseteq \mathscr{A}_1$, the variable $\eta = E(\xi_2|\mathscr{B}) \in L^2(\mathscr{A}_1)$ is the projection of ξ_2 on $L^2(\mathscr{A}_1)$:

$$\eta = E(\xi_2|\mathscr{A}_2)$$

and for $\xi = \xi_2$ this gives (1.5). $\qquad \square$

A trivial example of a σ-algebra which splits $\mathscr{A}_1$ and $\mathscr{A}_2$ is $\mathscr{B} = \mathscr{A}_1$. More generally, if $\mathscr{B}_0 \subseteq \mathscr{A}_1$ splits $\mathscr{A}_1$ and $\mathscr{A}_2$, then so does every σ-algebra $\mathscr{B}_1$ satisfying

$$\mathscr{B}_0 \subseteq \mathscr{B}_1 \subseteq \mathscr{A}_1, \tag{1.6}$$

since then we have, for $\xi \in L^2(\mathscr{A}_2)$

$$E(\xi|\mathscr{A}_1) = E(\xi|\mathscr{B}_0) = E(\xi|\mathscr{B}_1),$$

(cf. (1.5)). Moreover, if $\mathscr{B}_0$ is splitting for $\mathscr{A}_1$ and $\mathscr{A}_2$, then so is every σ-algebra $\mathscr{B} \supseteq \mathscr{B}_0$ of the form

$$\mathscr{B} = \mathscr{B}_1 \vee \mathscr{B}_2, \tag{1.7}$$

where $\mathscr{B}_1 \subseteq \mathscr{A}_1 \vee \mathscr{B}_0$ and $\mathscr{B}_2 \subseteq \mathscr{A}_2 \vee \mathscr{B}_0$.

This is because according to (1.3) and (1.6), the σ-algebra $\tilde{\mathscr{B}}_1 = \mathscr{B}_1 \vee \mathscr{B}_0$ splits the σ-algebras $\mathscr{A}_1 \vee \mathscr{B}_0$ and $\mathscr{A}_2 \vee \mathscr{B}_0$, and thus the σ-algebras $\tilde{\mathscr{A}}_1 = \mathscr{A}_1 \vee \tilde{\mathscr{B}}_1$ and $\tilde{\mathscr{A}}_2 = \mathscr{A}_2 \vee \tilde{\mathscr{B}}_1$; from the relation $\tilde{\mathscr{B}}_1 \subseteq \mathscr{B} \subseteq \tilde{\mathscr{A}}_2$ we conclude from (1.6) that $\mathscr{B}$ splits $\tilde{\mathscr{A}}_2 \supseteq \mathscr{A}_2$ and $\tilde{\mathscr{A}}_1 \supseteq \mathscr{A}_1$.

In connection with (1.6), (1.7) we remark that by no means all extensions $\mathscr{B} \supseteq \mathscr{B}_0$ of the σ-algebra $\mathscr{B}_0$ are splitting for $\mathscr{A}_1$ and $\mathscr{A}_2$ when $\mathscr{B}_0$ is splitting. One can show an easy example of this by noting that if $\mathscr{B}$ splits $\mathscr{A}_1$ and $\mathscr{A}_2$, then for any $\xi_1 \in L^2(\mathscr{A}_1)$ and $\xi_2 \in L^2(\mathscr{A}_2)$ the following must hold:

$$\xi_1 - E(\xi_1|\mathscr{B}) \perp \xi_2 - E(\xi_2|\mathscr{B}); \tag{1.8}$$

this is an immediate consequence of (1.2) if we write it in the form

$$E([\xi_1 - E(\xi_1|\mathscr{B})][\xi_2 - E(\xi_2|\mathscr{B})]|\mathscr{B}) = 0. \tag{1.9}$$

EXAMPLE. Let σ-algebras $\mathscr{A}_1$ and $\mathscr{A}_2$ be generated by independent Gaussian variables ξ_1 and ξ_2, and the σ-algebra $\mathscr{B}$ be generated by the variable $\eta = \xi_1 + \xi_2$. The trivial σ-algebra $\mathscr{B}_0$ splits $\mathscr{A}_1$ and $\mathscr{A}_2$, but $\mathscr{B} \supseteq \mathscr{B}_0$ does not have this property since (1.8) is not satisfied; we have

$$\eta_1 = E(\xi_1|\mathscr{B}) = c_1\eta, \qquad \eta_2 = E(\xi_2|\mathscr{B}) = c_2\eta,$$

$$\eta_1 + \eta_2 = \eta$$

and

$$(\xi_1 - \eta_1) + (\xi_2 - \eta_2) = \eta - (\eta_1 + \eta_2) = 0.$$

Let σ-algebras $\mathscr{B}_1, \mathscr{B}_2 \subseteq \mathscr{A}_1$ split $\mathscr{A}_1$ and $\mathscr{A}_2$. Then the intersection

$$\mathscr{B} = \mathscr{B}_1 \cap \mathscr{B}_2 \tag{1.10}$$

also splits $\mathscr{A}_1$ and $\mathscr{A}_2$, since we have (1.5) for $\mathscr{B} = \mathscr{B}_1, \mathscr{B}_2$ and thus the variable in this expression, $E(\xi|\mathscr{A}_1)$, which is measurable with respect to $\mathscr{B}_1$ and $\mathscr{B}_2$ is also measurable with respect to the σ-algebra $\mathscr{B} = \mathscr{B}_1 \cap \mathscr{B}_2$ contained in A_1, so that

$$E(\xi|\mathscr{A}_1) = E(\xi|\mathscr{B}).$$

Similarly, the intersection of an arbitrary number of splitting σ-algebras $\mathscr{B}$ contained in $\mathscr{A}_1$ splits $\mathscr{A}_1$ and $\mathscr{A}_2$. In general, if the condition that the $\mathscr{B} \subseteq \mathscr{A}_1$ is not met, then the intersection of the splitting σ-algebras $\mathscr{B}$ is not splitting. One can see this when the intersection $\mathscr{A}_1 \cap \mathscr{A}_2$ is not splitting for $\mathscr{A}_1$ and $\mathscr{A}_2$. (Here $\mathscr{B} = \mathscr{A}_1, \mathscr{A}_2$.)

Clearly, among splitting σ-algebras $\mathscr{B} \subseteq \mathscr{A}_1$, there is a minimal σ-algebra $\mathscr{B}_0 \subseteq \mathscr{A}_1$ which splits $\mathscr{A}_1$ and $\mathscr{A}_2$:

$$\mathscr{B}_0 \subseteq \mathscr{B} \tag{1.11}$$

for every σ-algebra $\mathscr{B} \subseteq \mathscr{A}_1$ which splits $\mathscr{A}_1$ and $\mathscr{A}_2$. ($\mathscr{B}_0$ can be obtained by taking the intersection of all splitting σ-algebras $\mathscr{B} \subseteq \mathscr{A}_1$.)

We point out the following easy result. Let $\mathscr{B}_n$, $n = 1, 2, \ldots$, be monotonically decreasing σ-algebras which split $\mathscr{A}_1$ and $\mathscr{A}_2$; they are not required to be contained in $\mathscr{A}_1$. Then their intersection

$$\mathscr{B} = \bigcap_n \mathscr{B}_n \tag{1.12}$$

also splits $\mathscr{A}_1$ and $\mathscr{A}_2$. One can see this by looking at the limit as $n \to \infty$ of equation (1.2) with $\mathscr{B} = \mathscr{B}_n$.

Each splitting σ-algebra $\mathscr{B}$ must necessarily contain the intersection $\mathscr{A}_0 = \mathscr{A}_1 \cap \mathscr{A}_2$:

$$\mathscr{A}_0 \subseteq \mathscr{B}, \tag{1.13}$$

since every random variable $\xi \in L^2(\mathscr{A}_0)$ satisfies (1.5)

$$\xi = E(\xi \,|\, \mathscr{A}_1) = E(\xi \,|\, \mathscr{B}),$$

which shows that ξ is measurable with respect to the σ-algebra $\mathscr{B}$.

2. Markov Random Processes

What is traditionally understood by the Markov property of a random process describing the behavior of some physical system under a time evolution $t \geq t_0$ states that when the present state of the system is known, its behavior in the future does not depend on the past. Starting with this property, we define a notion of "state" similar to that in classical mechanics.

Let the state of a system at time t be described by a random variable $\xi(t)$ in an appropriate phase space E. We denote by $\mathscr{A}(t_1, t_2)$ the σ-algebra of events generated by the variables $\xi(t)$, $t_1 \leq t \leq t_2$. The behavior of the system in the past and future can be described by the σ-algebras of events $\mathscr{A}(t_0, t)$ and $\mathscr{A}(t, \infty)$, and the Markov property expressed in terms of conditional probabilities by the equation

$$P(A \,|\, \xi(s), s \leq t) = P(A \,|\, \xi(t)), \qquad A \in \mathscr{A}(t, \infty); \tag{1.14}$$

this shows that for any t, the σ-algebra $\mathscr{A}(t)$ generated by $\xi(t)$ splits the "past," $\mathscr{A}(t_0, t)$, and the "future," $\mathscr{A}(t, \infty)$.

We note that on the basis of the general relation (1.13), $\mathscr{A}(t)$ is the minimal σ-algebra which splits $\mathscr{A}(t_0, t)$ and $\mathscr{A}(t, \infty)$, since

$$\mathscr{A}(t) \subseteq \mathscr{A}(t_0, t) \cap \mathscr{A}(t, \infty);$$

indeed,

$$\mathscr{A}(t) = \mathscr{A}(t_0, t) \cap \mathscr{A}(t, \infty). \tag{1.15}$$

Together with $\mathscr{A}(t)$ being splitting for $\mathscr{A}(t_0, t)$ and $\mathscr{A}(t, \infty)$, we also have σ-algebas

$$\mathscr{A}(t - \varepsilon, t + \varepsilon) = \mathscr{A}(t - \varepsilon, t) \vee \mathscr{A}(t, t + \varepsilon),$$

which are splitting for corresponding

$$\mathscr{A}(t_0, t + \varepsilon) = \mathscr{A}(t_0, t) \vee \mathscr{A}(t - \varepsilon, t + \varepsilon)$$

and

$$\mathscr{A}(t - \varepsilon, \infty) = \mathscr{A}(t, \infty) \vee \mathscr{A}(t - \varepsilon, t + \varepsilon)$$

(cf. (1.3), (1.7)). For all $\varepsilon > 0$ the σ-algebras $\mathscr{A}(t - \varepsilon, t + \varepsilon)$ split

$$\mathscr{A}_+(t_0, t) = \bigcap_{\varepsilon > 0} \mathscr{A}(t_0, t + \varepsilon)$$

and

$$\mathscr{A}_+(t, \infty) = \bigcap_{\varepsilon > 0} \mathscr{A}(t - \varepsilon, \infty);$$

as a consequence, the intersection

$$\mathscr{A}_+(t) = \bigcap_{\varepsilon > 0} \mathscr{A}(t - \varepsilon, t + \varepsilon) \tag{1.16}$$

also has this property and is the minimal σ-algebra which splits $\mathscr{A}_+(t_0, t)$ and $\mathscr{A}_+(t, \infty)$ and as in (1.15) we have

$$\mathscr{A}_+(t) = \mathscr{A}_+(t_0, t) \cap \mathscr{A}_+(t, \infty). \tag{1.17}$$

Here the σ-algebras $\mathscr{A}_+(t_0, t)$ and $\mathscr{A}_+(t, \infty)$ can be interpreted as the "total past" and "total future," and include events defined by the behavior of the process under consideration in infinitesimally small neighborhoods of the present time t.

A well-known example of a Markov process is the evolution of the real-valued random variables

$$\xi_1(t), \ldots, \xi_l(t), \qquad t \geq t_0, \tag{1.18}$$

which are equal to 0 for $t = t_0$ and satisfy (as generalized functions) a system of linear differential equations

$$\frac{d}{dt} \xi_k(t) = \sum_{j=1}^{l} a_{kj}(t)\xi_j(t) + \dot{\eta}_k(t), \qquad t > t_0, k = 1, \ldots, l, \tag{1.19}$$

with continuous coefficients and generalized random functions $\dot{\eta}_k(t)$ of "white noise" type:

$$(u, \dot{\eta}_k) = \int u(t)\eta_k(dt), \qquad u \in C_0(t_0, \infty),$$

where $\eta_k(dt)$ are Gaussian orthogonal measures with

$$E|\eta_k(dt)|^2 = b_k(t)\, dt,$$

characterized by locally integrable coefficients $b_k(t)$, $k = 1, \ldots, l$. Namely, if the state of the system at time t is given by the vector variable $\xi(t)$ with coordinates (1.18) in phase space $E = \mathbb{R}^l$, then the random function $\xi(t)$, $t \geq t_0$, will have the Markov property (1.14). This follows, for example, from the formula for the explicit solution of (1.19) with initial condition zero:

$$\xi(t) = \int_{t_0}^t w(t, s)\eta(ds), \qquad t \geq t_0, \tag{1.20}$$

where the $n \times n$ matrix $w(t, s)$, $t > s$, is the solution of the differential equation

$$\frac{d}{dt} w(t, s) = a(t) \cdot w(t, s),$$

with initial condition $w(s, s) = I$; here $a(t)$ is the $n \times n$ matrix whose elements are the coefficients $a_{kj}(t)$ in (1.19), I is the identity matrix, and $\eta(ds)$ appearing in (1.20) stands for the vector with coordinates $\eta_k(ds)$, $k = 1, \ldots, l$. We remark, for clarity, that

$$\xi_k(t) = \sum_{j=1}^l \int_{t_0}^t w_{kj}(t, s)\eta_j(ds), \qquad k = 1, \ldots, l,$$

where $w_{kj}(t, s)$ are the elements of the matrix $w(t, s)$. It is easy to show that the vector-valued function (1.20) satisfies the integral equation

$$\xi(t) = \int_{t_0}^t a(s)\xi(s)\, ds + \int_{t_0}^t \eta(ds), \qquad t \geq t_0;$$

The differential form of this equation is given for the components by (1.19). The Markov property (1.14) for $\xi(t)$, $t \geq t_0$, follows from the semi-group relation

$$w(u, s) = w(u, t) \cdot w(t, s), \qquad s \leq t \leq u,$$

which gives us the equality

$$\xi(u) = w(u, t)\xi(t) + \int_t^u w(u, s)\eta(ds).$$

It is obvious that for any $t \geq t_0$ the random variable $\xi(u) - w(u, t) \cdot \xi(t)$, $u \geq t$, defined for $\eta(\Delta)$, $\Delta \subseteq (t, \infty)$, does not depend on the variables $\xi(s)$,

$s \leq t$, defined according to (1.20) for $\eta(\Delta)$, $\Delta \subseteq (t_0, t)$; recall here that for Gaussian variables with mean zero, independence is equivalent to orthogonality and for Gaussian orthogonal measures $\eta_k(ds)$, $k = 1, \ldots, l$, the variables $\eta(\Delta)$, $\Delta \subseteq (t_0, t)$, and $\eta(\Delta)$, $\Delta \subseteq (t, \infty)$, are independent. The conditional probability distributions with respect to the σ-algebras $\mathscr{A}(t_0, t)$ are such that the variables $\xi(u)$, $u \geq t$, have Gaussian distributions with corresponding mean values

$$E(\xi(u) | \mathscr{A}(t_0, t)) = w(u, t)\xi(t)$$

and the same covariance matrices as the variables $\xi(u) - w(u, t)\xi(t)$, $u \geq t$.

We will look at an example which is typical of applications, arising from a system of the form (1.19). We consider the random motion of a particle on the real axis under a perturbing "white noise" influence; as a function of time $t > t_0$ its displacement $\xi(t)$ satisfies the linear differential equation

$$L\xi(t) \equiv \sum_{k=0}^{l} a_k(t) \frac{d^k}{dt^k} \xi(t) = \dot{\eta}(t), \tag{1.21}$$

with continuous coefficients $a_k(t)$ and "white noise" $\dot{\eta}(t)$ on the right-hand side. Let the leading coefficient in this expression be $a_l(t) = 1$. Then (1.21) is equivalent to a system (1.19) of the form

$$\frac{d}{dt} \xi_1(t) = \xi_2(t),$$

$$\vdots$$

$$\frac{d}{dt} \xi_{l-1}(t) = \xi_l(t),$$

$$\frac{d}{dt} \xi_l(t) = -\sum_{k=1}^{l} a_{k-1}(t)\xi_k(t) + \dot{\eta}(t),$$

where $\xi_1(t) = \xi(t)$. It is apparent that for $l > 1$, the individual random function $\xi(t)$ does not have the Markov property (1.14) however the l-dimensional vector-valued function with components

$$\xi_k(t) = \frac{d^{k-1}}{dt^{k-1}} \xi(t), \qquad k = 1, \ldots, l, \tag{1.22}$$

does have it. All these derivatives are ordinary functions with values in $L^2(\Omega, \mathscr{A}, P)$. For $t_1 \leq t \leq t_2$, the components in (1.22) are measurable with respect to the σ-algebra $\mathscr{A}(t_1, t_2)$ generated by variables $\xi(t), t_1 \leq t \leq t_2$, with $t_2 > t_1$, and we claim that for any $\varepsilon > 0$, the σ-algebra $\mathscr{A}(t - \varepsilon, t + \varepsilon)$ splits $\mathscr{A}(t_0, t)$ and $\mathscr{A}(t, \infty)$ for every $t > t_0$.

We get this result from a definition of *Markov behavior* applicable to generalized random functions

$$(u, \xi), \qquad u \in C_0^{\infty}(t_0, \infty);$$

here values are not defined for separate points t, but associated with each interval (t_1, t_2) there is a σ-algebra of events $\mathscr{A}(t_1, t_2)$ generated by variables (u, ξ) with Supp $u \subseteq (t_1, t_2)$.

Specifically, our notion of Markov behavior of a generalized random function will be that for every moment $t > t_0$ and for all sufficiently small $\varepsilon > 0$, the σ-algebras $\mathscr{A}(t - \varepsilon, t + \varepsilon)$ split "the past," $\mathscr{A}(t_0, t)$, and "the future," $\mathscr{A}(t, \infty)$.

We note that this property is strictly stronger than that defined by (1.16)— that $\mathscr{A}_+(t)$ be splitting for $\mathscr{A}(t_0, t)$ and $\mathscr{A}(t, \infty)$. For example, we have the trivial σ-algebra $\mathscr{A}_+(t)$ splitting independent $\mathscr{A}(t_0, t)$ and $\mathscr{A}(t, \infty)$ when the generalized function is the derivative of white noise,

$$(u, \xi) = \int u'(t)\eta(dt), \qquad u \in C_0^\infty(t_0, \infty),$$

where $\eta(dt)$ is the Gaussian orthogonal measure, $E|\eta(dt)|^2 = dt$; but this generalized random function does not have Markov behavior. In fact, in the case of Markov behavior, the σ-algebra $\mathscr{B} = \mathscr{A}(t - \varepsilon, t + \varepsilon)$ splits $\mathscr{A}_1 = \mathscr{A}(t_0, t + \varepsilon)$ and $\mathscr{A}_2 = \mathscr{A}(t - \varepsilon, \infty)$, which directly implies (1.8); but this condition is violated by our example, as is easily checked for the variables

$$\xi_1 = \eta(t - \varepsilon, t + \varepsilon) - \eta(t - 3\varepsilon, t - \varepsilon)$$

and

$$\xi_2 = \eta(t + \varepsilon, t + 3\varepsilon) - \eta(t - \varepsilon, t + \varepsilon);$$

they are orthogonal to the variables (u, ξ), Supp $u \subseteq (t - \varepsilon, t + \varepsilon)$, generating the σ-algebra $\mathscr{A}(t - \varepsilon, t + \varepsilon)$ but are not orthogonal to each other. (A similar but more detailed example will be analyzed in Chapter 3, §1.2.)

3. Random Fields; Markov Property

Let certain σ-algebras of events $\mathscr{A}(S)$ be connected with domains $S \subseteq T$, T a locally-compact metric space. We consider all open domains $S \subseteq T$ and call the family $\mathscr{A}(S)$, $S \subseteq T$, a *random field* if it has the following additive property:

$$\mathscr{A}(S' \cup S'') = \mathscr{A}(S') \vee \mathscr{A}(S''). \tag{1.23}$$

A random field is *continuous* if

$$\mathscr{A}(S) = \bigvee_n \mathscr{A}(S^{(n)}), \qquad S = \bigcup_n S^{(n)} \tag{1.24}$$

for a monotonically increasing sequence of open domains $S^{(n)}$, $n = 1, 2, \ldots$.

It is immediately apparent that the σ-algebras of events $\mathscr{A}(S)$ forming a random field depend monotonically on $S \subseteq T$,

$$\mathscr{A}(S') \subseteq \mathscr{A}(S''), \qquad S' \subseteq S'', \tag{1.25}$$

because using the additive property for $S'' = S' \cup S''$ we have

$$\mathscr{A}(S'') = \mathscr{A}(S') \vee \mathscr{A}(S'').$$

EXAMPLE (Random Functions). Let $\xi(t), t \in T$, be a random function and $\mathscr{A}(S)$ be the σ-algebra of events generated by variables $\xi(t)$, $t \in S$. It is obvious that the collection $\mathscr{A}(S)$, $S \subseteq T$, forms a continuous random field.

EXAMPLE (Generalized Random Functions). Let $T \subseteq \mathbb{R}^d$ be a domain in d-dimensional Euclidean space and let (u, ξ), $u \in C_0^\infty(T)$, be a generalized random function. For an open domain $S \subseteq T$, let $\mathscr{A}(S)$ be the σ-algebra of events generated by variables (u, ξ) with Supp $u \subseteq S$. For any open domains S_k, $k = 1, 2, \ldots$, the following equality holds:

$$\mathscr{A}(S) = \bigvee_k \mathscr{A}(S_k), \qquad S = \bigcup_k S_k.$$

In fact, for any function $u \in C_0^\infty(T)$ with support Supp $u \subseteq S$, there is a finite open covering of the compact set Supp u, Supp $u \subseteq \bigcup_{j=1}^n S_{k_j}$, such that by taking an appropriate partition of unity

$$1 = \sum_{j=1}^n u_j(t), \qquad t \in \text{Supp } u,$$

consisting of components $u_j \in C_0^\infty(T)$ having support Supp $u_j \subseteq S_{k_j}$ we get that

$$u = \sum_{j=1}^n u \cdot u_j, \qquad (u, \xi) = \sum_{j=1}^n (u \cdot u_j, \xi),$$

with the components $(u \cdot u_j, \xi)$ measurable with respect to $\mathscr{A}(S_{k_j})$, $j = 1, \ldots, n$. Thus the family $\mathscr{A}(S)$, $S \subseteq T$, forms a continuous random field.

EXAMPLE (Stochastic Measures). Let $\eta(dt)$, $t \in T$, be a stochastic orthogonal measure in a domain $T \subseteq \mathbb{R}^d$, with

$$E|\eta(dt)|^2 = \mu(dt).$$

We let $\mathscr{A}(S)$ denote the σ-algebra of events generated by all variables $\eta(\Delta)$, $\Delta \subseteq S \subseteq T$. The family $\mathscr{A}(S)$, $S \subseteq T$, forms a continuous random field, since for $\Delta \subseteq S' \cup S''$

$$\eta(\Delta) = \eta(\Delta \cap S') + \eta(\Delta \backslash S'),$$

where $\Delta \backslash S'$ is contained in S'', and for a monotone increasing sequence of domains $S^{(n)}$ with $\Delta \subseteq \bigcup S^{(n)}$ we have

$$\|\eta(\Delta) - \eta(\Delta \cap S^{(n)})\|^2 = \mu(\Delta \backslash S^{(n)}) \to 0.$$

In conjunction with this random field we define $H(S)$, the closed linear span of the variables $\eta(\Delta)$, $\Delta \subseteq S$, in the spaces $L^2(\Omega, \mathscr{A}, P)$. Clearly,

$$H(S' \cup S'') = H(S') \vee H(S''),$$

where the symbol $\vee$, used in this context, denotes the closed linear span of the indicated spaces. Moreover,

$$H(S) = \bigvee H(S^{(n)}),$$

when $S = \bigcup_n S^{(n)}$ (cf. (1.23) and (1.24)).

We remark that in all our examples of random fields $\mathscr{A}(S)$, $S \subseteq T$, it turns out that

$$\mathscr{A}(S) = \bigvee_{S' \subseteq S} \mathscr{A}(S')$$

is generated by events determined by all possible relatively compact domains $S' \subseteq S$.

Now we consider an arbitrary random field $\mathscr{A}(S)$, $S \subseteq T$, S open.

We will say that the domain S splits the domains S_1 and S_2 if the σ-algebra $\mathscr{A}(S)$ splits $\mathscr{A}(S_1)$ and $\mathscr{A}(S_2)$. We point out two properties of splitting domains which we will use frequently.

If the domain S splits S_1 and S_2, then it also splits $\tilde{S}_1 = S_1 \cup S$ and $\tilde{S}_2 = S_2 \cup S$, since

$$\mathscr{A}(\tilde{S}_1) = \mathscr{A}(S_1) \vee \mathscr{A}(S), \qquad \mathscr{A}(\tilde{S}_2) = \mathscr{A}(S_2) \vee \mathscr{A}(S)$$

see (1.3).

If S_0 splits S_1 and S_2, then so does any domain $S \supseteq S_0$ satisfying the condition

$$S \subseteq (S_1 \cup S_0 \cup S_2),$$

since

$$S = (S_1 \cap S) \cup S_0 \cup (S_2 \cap S)$$

and

$$\mathscr{A}(S) = \mathscr{A}(S_1 \cap S) \vee \mathscr{A}(S_0) \vee \mathscr{A}(S_2 \cap S),$$

and by the monotone property

$$\mathscr{A}(S_1 \cap S) \subseteq \mathscr{A}(S_1), \qquad \mathscr{A}(S_2 \cap S) \subseteq \mathscr{A}(S_2);$$

now apply (1.6) and (1.7).

When looking at a closed set $\Gamma \subseteq T$ we will say that it splits domains S_1 and S_2 if these domains are split by every sufficiently small ε-neighborhood Γ^ε of the set Γ.

We let $\mathscr{G}$ be a system of open domains $S \subseteq T$, and for each S we define the triple

$$S_1 = S, \Gamma, S_2 = T \backslash \bar{S}, \tag{1.26}$$

where Γ is a closed set containing the (topological) boundary ∂S of the domain S,

$$\Gamma \supseteq \partial S,$$

and $\bar{S} = S \cup \partial S$ is the closure of the set S in the space T; we will call $S_2 = T\backslash\bar{S}$ the *complementary domain* for $S_1 = S$ and the set Γ a *boundary* between S_1 and S_2.

We call a random field $\mathscr{A}(S)$, $S \subseteq T$, *Markov with respect to the system* $\mathscr{G}$, if for every domain $S \subseteq \mathscr{G}$ the boundary Γ splits $S_1 = S$ and $S_2 = T\backslash\bar{S}$, in other words, if the σ-algebras

$$\mathscr{A}(S_1), \quad \mathscr{A}(\Gamma^\varepsilon), \quad \mathscr{A}(S_2) \tag{1.27}$$

form a Markov sequence for all small neighborhoods Γ^ε of the boundary Γ between the domains S_1 and S_2.

We recall here that if Γ^ε splits complementary domains S_1 and S_2, then any neighborhood $\Gamma^\delta \supseteq \Gamma^\varepsilon$ also has this property since

$$\Gamma^\delta = (S_1 \cap \Gamma^\delta) \cup \Gamma^\varepsilon \cup (S_2 \cap \Gamma^\delta).$$

Let us turn to an investigation of the Markov property.

If Γ^ε splits S_1 and S_2, then Γ^ε splits the domains $S_1 \cup \Gamma^\varepsilon$ and $S_2 \cup \Gamma^\varepsilon$—i.e., the sequence

$$\mathscr{A}(S_1 \cup \Gamma^\varepsilon), \quad \mathscr{A}(\Gamma^\varepsilon), \quad \mathscr{A}(S_2 \cup \Gamma^\varepsilon)$$

is Markov.

Let the random field be defined on all sufficiently small ε-neighborhoods S^ε of a set $S \subseteq T$; we set

$$\mathscr{A}_+(S) = \bigcap_{\varepsilon > 0} \mathscr{A}(S^\varepsilon).$$

If S is an open domain and Γ is a closed set containing the topological boundary ∂S, then $(S \cup \Gamma)^\varepsilon = S \cup \Gamma^\varepsilon$ and in particular we have for the triple (1.26),

$$(S_1 \cup \Gamma)^\varepsilon = S_1 \cup \Gamma^\varepsilon, \qquad (S_2 \cup \Gamma)^\varepsilon = S_2 \cup \Gamma^\varepsilon,$$

since $\partial S_2 \subseteq \partial S_1 \subseteq \Gamma$. We know that if σ-algebras $\mathscr{A}(\Gamma^\varepsilon)$ split

$$\mathscr{A}_+(S_1 \cup \Gamma) \subseteq \mathscr{A}(S_1 \cup \Gamma^\varepsilon) \quad \text{and} \quad \mathscr{A}_+(S_2 \cup \Gamma) \subseteq \mathscr{A}(S_2 \cup \Gamma^\varepsilon),$$

then the intersection

$$\mathscr{A}_+(\Gamma) = \bigcap \mathscr{A}(\Gamma^\varepsilon)$$

splits the σ-algebras $\mathscr{A}_+(S_1 \cup \Gamma)$ and $\mathscr{A}_+(S_2 \cup \Gamma)$; moreover $\mathscr{A}_+(\Gamma)$ is the minimal σ-algebra having the property

$$\mathscr{A}_+(\Gamma) = \mathscr{A}_+(S_1 \cup \Gamma) \cap \mathscr{A}_+(S_2 \cup \Gamma) \tag{1.28}$$

(cf. (1.13)), and of course $\mathscr{A}_+(\Gamma)$ splits the σ-algebras $\mathscr{A}(S_1)$ and $\mathscr{A}(S_2)$. Hence the sequence

$$\mathscr{A}(S_1), \quad \mathscr{A}_+(\Gamma), \quad \mathscr{A}(S_2) \tag{1.29}$$

is Markov.

For any domain $S \in \mathcal{G}$, suppose the corresponding sequence (1.29) is Markov. Will the related random field be Markov?

The answer to this question is yes, provided that for neighborhoods Γ^ε the following equality is true:

$$\mathcal{A}(\Gamma^\varepsilon) = \mathcal{A}(S_1 \cap \Gamma^\varepsilon) \vee \mathcal{A}_+(\Gamma) \vee \mathcal{A}(S_2 \cap \Gamma^\varepsilon). \tag{1.30}$$

This follows because if $\mathcal{A}_+(\Gamma)$ splits the σ-algebras $\mathcal{A}(S_1)$ and $\mathcal{A}(S_2)$, then so does the σ-algebra $\mathcal{A}(\Gamma^\varepsilon)$ because the representation (1.30) gives

$$\mathcal{A}(S_1 \cap \Gamma^\varepsilon) \subseteq \mathcal{A}(S_1), \qquad \mathcal{A}(S_2 \cap \Gamma^\varepsilon) \subseteq \mathcal{A}(S_2),$$

and (1.7) applies.

EXAMPLE (White Noise). Consider a generalized random function on a domain $T \subseteq \mathbb{R}^d$ of the form

$$(u, \xi) = \int u(t)\eta(dt), \qquad u \in C_0^\infty(T),$$

where $\eta(dt)$ is the Gaussian orthogonal measure

$$E|\eta(dt)|^2 = dt.$$

For an open domain $S \subseteq T$, the σ-algebra $\mathcal{A}(S)$ generated by variables (u, ξ) with support $\text{Supp } u \subseteq S$ coincides with the σ-algebra $\mathcal{B}(S)$ generated by the subspace

$$H(S) \subseteq L^2(\Omega, \mathcal{A}, P)$$

of Gaussian variables of the form

$$\eta = \int u(t)\eta(dt), \qquad u \in L^2(S),$$

where $L^2(S)$ denotes† the subspace of functions equal to 0 outside S in the standard Hilbert space $L^2(T)$ of real-valued square-integrable functions $u = u(t)$, $t \in T$. The subspaces $H(S)$ and $L^2(S)$ are unitarily isomorphic:

$$E\eta_1\eta_2 = \int u_1(t)u_2(t)\, dt$$

for variables $\eta = \eta_1$, $\eta_2 \in H(S)$ and $u = u_1$, $u_2 \in L^2(S)$ as indicated above. Therefore for any measurable $S \subseteq T$ we have

$$\mathcal{A}_+(S) = \bigcap_{\varepsilon > 0} B(S^\varepsilon) = \mathcal{B}(\bar{S}),$$

since the σ-algebra $\mathcal{B}(\bar{S})$ is generated by the subspace

$$H(\bar{S}) = \bigcap_{\varepsilon > 0} H(S^\varepsilon),$$

---

† We hope this change of notation for L^2-type spaces does not cause confusion.

which is unitarily isomorphic to the subspace

$$L^2(\bar{S}) = \bigcap_{\varepsilon > 0} L^2(S^\varepsilon).$$

We consider the collection of all open domains $S \subseteq T$ with boundary $\Gamma = \partial S$.

The subspaces

$$H(S_1), H(\Gamma), H(S_2)$$

are orthogonal for disjoint

$$S_1 = S, \qquad \Gamma = \partial S, \qquad S_2 = T \backslash \bar{S}$$

and thus the σ-algebras generated by them

$$\mathscr{A}(S_1) = \mathscr{B}(S_1), \qquad \mathscr{A}_+(\Gamma) = \mathscr{B}(\Gamma), \qquad \mathscr{A}(S_2) = \mathscr{B}(S_2)$$

are independent. In particular, the corresponding sequence (1.29) is Markov for any open domain $S \subseteq T$. It is clear that (1.30) holds since

$$\Gamma^\varepsilon = (S_1 \cap \Gamma^\varepsilon) \cup \Gamma \cup (S_2 \cap \Gamma^\varepsilon)$$

and

$$H(\Gamma^\varepsilon) = H(S_1 \cap \Gamma^\varepsilon) \vee H(\Gamma) \vee H(S_2 \cap \Gamma^\varepsilon).$$

In this way, the "white noise" random field $\mathscr{A}(S)$, $S \subseteq T$, is Markov with respect to the system of all open domains $S \subseteq T$ with boundary $\Gamma = \partial S$.

One should note that the answer to the question posed in connection with the Markov sequence (1.29) is negative in the general case. For instance, for the derivative of "white noise," the corresponding σ-algebras

$$\mathscr{A}(S_1) \subseteq \mathscr{B}(S_1), \qquad \mathscr{A}_+(\Gamma) \subseteq \mathscr{B}(\Gamma), \qquad \mathscr{A}(S_2) \subseteq \mathscr{B}(S_2)$$

are independent, but the sequence (1.27) need not be Markov (in this context, see the example on p. 62).

EXAMPLE (Stochastic Measures with Independent Values). We consider the random field $\mathscr{A}(S)$, $S \subseteq T$, which is generated by a stochastic orthogonal measure $\eta(dt)$ in a domain $T \subseteq \mathbb{R}^d$,

$$E|\eta(dt)|^2 = \mu(dt);$$

see the earlier example in this section. We assume that for any disjoint sets $\Delta_1, \ldots, \Delta_n \subseteq T$ the values $\eta(\Delta_1), \ldots, \eta(\Delta_n)$ are independent random variables. For instance, one could consider variables

$$\eta(\Delta) = \xi(\Delta) - E\xi(\Delta),$$

where $\xi(\Delta)$, $\Delta \subseteq T$, is a Poisson point process in the domain $T \subseteq \mathbb{R}^d$ ($\xi(\Delta)$ is the number of random points falling in the set Δ when the points have a

Poisson distribution in the domain T). Recall that $\mathscr{A}(S)$ is the σ-algebra of events generated by all possible variables $\eta(\Delta)$, $\Delta \subseteq S$, S open, and it is convenient to use this definition for any measurable set $S \subseteq T$. For any open region $S \subseteq T$ with boundary $\Gamma = \partial S$ and $S_1 = S$, $S_2 = T \backslash \bar{S}$, the σ-algebras

$$\mathscr{A}(S_1), \mathscr{A}(\Gamma), \mathscr{A}(S_2)$$

are independent and in particular, the σ-algebra $\mathscr{A}(\Gamma)$ splits $\mathscr{A}(S_1)$ and $\mathscr{A}(S_2)$. From the additivity property of stochastic measures $\eta(dt)$ it follows that

$$\mathscr{A}(\Gamma^\varepsilon) = \mathscr{A}(S_1 \cap \Gamma^\varepsilon) \vee \mathscr{A}(\Gamma) \vee \mathscr{A}(S_2 \cap \Gamma^\varepsilon),$$

since

$$\Gamma^\varepsilon = (S_1 \cap \Gamma^\varepsilon) \cup \Gamma \cup (S_2 \cap \Gamma^\varepsilon);$$

hence together with $\mathscr{A}(\Gamma)$ the σ-algebra $\mathscr{A}(\Gamma^\varepsilon)$, $\varepsilon > 0$, splits $\mathscr{A}(S_1)$ and $\mathscr{A}(S_2)$; that is, the random field we are considering is Markov with respect to the system of all open domains $S \subseteq T$. $\qquad \square$

Now we consider an arbitrary random field $\mathscr{A}(S)$, $S \subseteq T$. We call the random field *Markov* if it is Markov with respect to a complete system $\mathscr{G}$ of open domains $S \subseteq T$ and the canonical triple for each S consists of $S_1 = S$, $S_2 = T \backslash \bar{S}$, and $\Gamma = \partial S$, the topological boundary between S_1 and S_2. A *complete* system $\mathscr{G}$ is one having the following properties. It contains all open domains which are relatively compact or have compact complements. It contains all sufficiently small ε-neighborhoods of the boundary $\Gamma = \partial S$, $S \in \mathscr{G}$. Moreover, if S' and S'' are in $\mathscr{G}$, their union is also in $\mathscr{G}$, and if $S_1 = S$ is in $\mathscr{G}$ then $S_2 = T \backslash \bar{S}$ is in $\mathscr{G}$.

Let our random field be Markov. For each domain $S \in \mathscr{G}$, let

$$\Gamma^\varepsilon_- = S \cap \Gamma^\varepsilon$$

be a sufficiently small one-sided neighborhood of the boundary $\Gamma = \partial S$; we will show that Γ^ε_- splits S_1 and S_2, i.e., the sequence

$$\mathscr{A}(S_1), \mathscr{A}(\Gamma^\varepsilon_-), \mathscr{A}(S_2) \qquad (1.31)$$

is Markov, and in turn, if every sequence of the form (1.31) is Markov then the random field is Markov. Let

$$\tilde{S} = S \backslash \overline{\Gamma^\delta}, \qquad \overline{\Gamma^\delta} \subseteq \Gamma^\varepsilon.$$

The domain $\tilde{S}$ has boundary $\tilde{\Gamma} = \partial \tilde{S}$ which, together with a small enough neighborhood, is contained in the neighborhood $\Gamma^\varepsilon_- = S \cap \Gamma^\varepsilon$, so that Γ^ε_- splits $\tilde{S}_1 = \tilde{S}$ and $\tilde{S}_2 = T \backslash \tilde{S}$ and thus also splits $S_1 = \tilde{S}_1 \cup \Gamma^\varepsilon_-$ and $S_2 \cup \Gamma^\varepsilon \subseteq \tilde{S}_2 \cup \Gamma^\varepsilon_-$. In turn, if the one-sided neighborhood Γ^ε_- splits S_1 and $S_2 \cup \Gamma^\varepsilon$ then this same property holds for the complete neighborhood $\Gamma^\varepsilon \supseteq \Gamma^\varepsilon_-$, contained in the domain $S_2 \cup \Gamma^\varepsilon$.

As a consequence we have that *for a Markov field, the σ-algebra*

$$\mathscr{A}_+(\Gamma_-) = \bigcap \mathscr{A}(\Gamma_-^\varepsilon)$$

splits $\mathscr{A}(S_1)$ *and* $\mathscr{A}(S_2) \subseteq \mathscr{A}_+(S_2 \cup \Gamma)$.

For a neighborhood Γ^ε of the boundary $\Gamma = \partial S$, let the following be true:

$$\mathscr{A}(\Gamma^\varepsilon) = \mathscr{A}(S_1 \cap \Gamma^\varepsilon) \vee \mathscr{A}(S_2 \cap \Gamma^\varepsilon), \tag{1.32}$$

keeping in mind that $S_1 = S$ and $S_2 = T \backslash \bar{S}$. For instance, this will be true if the random field is generated by a continuous random function $\xi(t)$, $t \in T$—more precisely, if the σ-algebras $\mathscr{A}(S)$, $S \subseteq T$, are generated by the corresponding random variables $\xi(t)$, $t \in S$.

Under condition (1.32), in addition to the σ-algebra $\mathscr{A}_+(\Gamma_-) \subseteq \mathscr{A}(S_1)$ splitting $\mathscr{A}(S_1)$, $\mathscr{A}(S_2)$ we have that the σ-algebra $\mathscr{A}(\Gamma^\varepsilon) \supseteq \mathscr{A}_+(\Gamma_-)$ also splits $\mathscr{A}(S_1)$, $\mathscr{A}(S_2)$ (cf. 1.7)).

In sum, we have shown the following proposition:

Under condition (1.32), the random field $\mathscr{A}(S)$, $S \subseteq T$, *being Markov is equivalent to the condition that for any triple (1.26) the sequence*

$$\mathscr{A}(S_1), \mathscr{A}_+(\Gamma_-), \mathscr{A}(S_2) \tag{1.33}$$

is Markov.

Note that

$$\mathscr{A}_+(\Gamma_-) = \mathscr{A}_+(\Gamma), \qquad \Gamma = \partial S, \tag{1.34}$$

when the Markov field is *exterior continuous*—that is,

$$\mathscr{A}_+(S) = \bigcap \mathscr{A}(S^\varepsilon) = \mathscr{A}(S);$$

this can be interpreted as continuity of renewal of the random field for the expanding domains S^ε, $\varepsilon > 0$. In fact, since it is the minimal σ-algebra splitting $\mathscr{A}_+(S_1) = \mathscr{A}_+(S_1 \cup \Gamma)$ and $\mathscr{A}_+(S_2 \cup \Gamma)$, $\mathscr{A}_+(\Gamma)$ coincides with the σ-algebra $\mathscr{A}_+(\Gamma_-) \subseteq \mathscr{A}_+(\Gamma)$, which splits $\mathscr{A}_+(S_1) = \mathscr{A}(S_1)$ and $\mathscr{A}_+(S_2 \cup \Gamma)$.

4. Transformations of Distributions which Preserve the Markov Property. Additive Functionals

Let $\mathscr{A}(S)$, $S \subseteq T$, be a random field formed by σ-algebras of events $\mathscr{A}(S) \subseteq \mathscr{A}$ in a probability space $(\Omega, \mathscr{A}, P)$. We defined the Markov property with respect to a system of domains $S \in \mathscr{G}$ with boundaries $\Gamma \supseteq \partial S$ in such a way that for a probability P on a σ-algebra $\mathscr{A}(T)$, the Markov property implies that for any $S_1 = S$, $S_2 = T \backslash \bar{S}$, and sufficiently small neighborhood Γ^ε the following equation holds:

$$E(\xi | \mathscr{A}_1, P) = E(\xi | \mathscr{B}, P), \qquad \xi \in L^1(\Omega, \mathscr{A}_2, P),$$

where $E(\xi|\cdot, P)$ stands for conditional expectation for the probability measure P and

$$\mathcal{A}_1 = \mathcal{A}(S_1 \cup \Gamma^\varepsilon), \qquad \mathcal{B} = \mathcal{A}(\Gamma^\varepsilon), \qquad \mathcal{A}_2 = \mathcal{A}(S_2 \cup \Gamma^\varepsilon).$$

Let Q be another probability measure on the space $(\Omega, \mathcal{A})$, having the form

$$Q(A) = \int_A p(\omega)P(d\omega), \qquad A \in \mathcal{A}, \tag{1.35}$$

where the density

$$p = Q(d\omega)/P(d\omega)$$

admits a decomposition

$$p = p_1 \cdot p_2, \tag{1.36}$$

with the factors p_1 and p_2 measurable with respect to the corresponding $\mathcal{A}_1 = \mathcal{A}(S_1 \cup \Gamma^\varepsilon)$ and $\mathcal{A}_2 = \mathcal{A}(S_2 \cup \Gamma^\varepsilon)$. We call the variable $p \in L^1$ $(\Omega, \mathcal{A}, P)$ *multiplicative* with respect to the system of domains $S \in \mathcal{G}$ if it has the above property and moreover, one of the factors in (1.36) is an L^1-function, say,

$$p_2 \in L^1(\Omega, \mathcal{A}_2, P).$$

The following proposition is true:

A transformation (1.35) with multiplicative density preserves the Markov property.

We will show this after first specifying certain properties of conditional expectations.

Consider an arbitrary σ-algebra $\mathcal{B} \subseteq \mathcal{A}$ and variable η, measurable with respect to $\mathcal{B}$. Let the variable ξ and the product $\xi\eta$ be integrable,

$$\xi, \xi\eta \in L^1(\Omega, \mathcal{A}, P).$$

Then we have the equality:

$$E(\xi\eta|\mathcal{B}) = E(\xi|\mathcal{B}) \cdot \eta. \tag{1.37}$$

To prove this we look at the functions

$$\xi_k = \begin{cases} \xi, & |\xi| \le k, \\ 0, & \text{else.} \end{cases}$$

$$\eta_k = \begin{cases} \eta, & |\eta| \le k, \\ 0, & \text{else,} \end{cases} \qquad k = 1, 2, \dots.$$

Then the sequences ξ_k and $\xi_k \eta_k$ converge in L^1 to ξ and $\xi\eta$ respectively. But the operator $E(\cdot | \mathscr{B})$ is continuous in L^1, and taking the limit in the equation

$$E(\xi_k \eta_k | \mathscr{B}) = E(\xi_k | \mathscr{B}) \eta_k$$

we get (1.37).

Let Q be a transformation (1.35) of the probability measure P with density $p = Q(d\omega)/P(d\omega)$ and let

$$q = E(p | \mathscr{B}, P).$$

We have $q(\omega) \neq 0$ for Q-almost all $\omega \in \Omega$ and

$$E(\xi | \mathscr{B}, Q) = \frac{1}{q} E(p\xi | \mathscr{B}, P). \tag{1.38}$$

Indeed, the variable q is the density of Q with respect to the σ-algebra $\mathscr{B} \subseteq \mathscr{A}$:

$$Q(B) = \int_B p(\omega) P(d\omega) = \int_B q(\omega) P(d\omega), \qquad B \in \mathscr{B},$$

from which it follows that

$$\int_B \frac{1}{q} E(p\xi | \mathscr{B}, P) Q(d\omega) = \int_B E(p\xi | \mathscr{B}, P) P(d\omega)$$

$$= \int_B p\xi(\omega) P(d\omega) = \int_B \xi(\omega) Q(d\omega), \qquad B \in \mathscr{B}.$$

Now we prove that the transformation (1.35) preserves the Markov property.

Let the random field be Markov under the distribution P. We must establish that if a variable ξ is measurable with respect to the σ-algebra $\mathscr{A}_2$ then if $\mathscr{B} = \mathscr{A}(\Gamma^\varepsilon)$ the equation

$$E(\xi | \mathscr{A}_1, Q) = E(\xi | \mathscr{B}, Q)$$

holds (cf. (1.5)). By the general formula (1.38)

$$E(\xi | \mathscr{A}_1, Q) = \frac{E(p\xi | \mathscr{A}_1, P)}{E(p | \mathscr{A}_1, P)},$$

and by the multiplicative property (1.36), the general formula (1.37), and the Markov behavior for the distribution P we have

$$E(p | \mathscr{A}_1, P) = p_1 E(p_2 | \mathscr{A}_1, P) = p_1 E(p_2 | \mathscr{B}, P),$$

$$E(p\xi | \mathscr{A}_1, P) = p_1 E(p_2 \xi | \mathscr{A}_1, P) = p_1 E(p_2 \xi | \mathscr{B}, P)$$

Clearly,

$$E(\xi | \mathscr{A}_1, Q) = \frac{E(p_2 \xi | \mathscr{B}, P)}{E(p_2 | \mathscr{B}, P)}$$

is a variable measurable with respect to the σ-algebra $\mathscr{B}$ and therefore it coincides with the conditional expectation $E(\xi \,|\, \mathscr{B}, Q)$, which is what we wished to show.

An *additive functional* on the random field $\mathscr{A}(S)$, $S \subseteq T$, is a real-valued random function $\xi(S)$, $S \subseteq T$, defined on the same system of open domains $S \subseteq T$ as the random field and having the following property:

$$\xi(S_1 \cup S_2) = \xi(S_1) + \xi(S_2) - \xi(S_1 \cap S_2)$$

for any S_1, S_2. Under the further condition

$$E \exp[\xi(S)] < \infty, \qquad S \subseteq T, \tag{1.39}$$

the additive functional gives us a multiplicative probability density

$$p = \frac{1}{E \exp[\xi(T)]} \exp[\xi(T)].$$

As an example of a Gaussian Markov field we considered above Gaussian white noise

$$(u, \dot{\eta}) = \int u(t)\eta(dt), \qquad u \in C_0^\infty(T),$$

in the domain $T \subseteq \mathbb{R}^d$ with $\eta(dt)$ the Gaussian orthogonal measure, $E|\eta(dt)|^2 = dt$. This example raises the question of what we can represent as an additive functional on this Markov field.

We restrict our attention to additive functionals with finite second moment

$$E|\xi(S)|^2 < \infty, \qquad S \subseteq T;$$

we do not require that they satisfy condition (1.39) but instead require that they be continuous in the sense that for a monotonically shrinking sequence of regions $S^{(n)}$, $n = 1, 2, \ldots$,

$$\xi(S^{(n)}) \to 0 \quad \text{for} \quad \bigcap_n \mathscr{A}(S^{(n)}) = \varnothing.$$

Actually, what we need is just the property that if the boundary $\Gamma = \partial S$ of the domain $S \subseteq T$ has measure 0 then

$$\xi(\Gamma^\varepsilon) \to 0 \quad \text{as } \varepsilon \to 0.$$

We partition the domain T, using disjoint open domains $S_1, \ldots, S_n$ with corresponding boundaries $\Gamma = \partial S$ of measure 0 so that $T = \bigcup_{k=1}^n \bar{S}_k$; this gives us

$$\xi(T) = \sum_{k=1}^n \xi(S_k). \tag{1.40}$$

We turn to the space $H_p(S)$ formed by pth degree Hermite polynomials of the variables $(u, \dot{\eta})$, Supp $u \subseteq S$. As we have seen, each variable $\eta \in H_p(S)$ is given by a (p times) *iterated Ito integral*:

$$\eta = \int \cdots \int_{S \times \cdots \times S} \varphi(t_1, \ldots, t_p) \eta(dt_1) \cdots \eta(dt_p);$$

moreover,

$$\|\eta\|^2 = p! \int \cdots \int_{S \times \cdots \times S} |\varphi(t_1, \ldots, t_p)|^2 \, dt_1 \cdots dt_p.$$

We denote by $\eta(S)$ the projection of the variable $\xi(S)$ on the subspace $H_p(T)$; it coincides with the projection on $H_p(S)$. On the basis of this, it follows from (1.40) that

$$\eta(T) = \sum_{k=1}^{n} \oplus \eta(S_k),$$

where, under

$$\eta(T) = \int \cdots \int_{T \times \cdots \times T} \varphi(t_1, \ldots, t_p) \eta(dt_1) \cdots \eta(dt_p),$$

we have

$$\eta(S_k) = \int \cdots \int_{S_k \times \cdots \times S_k} \varphi(t_1, \ldots, t_p) \eta(dt_1) \cdots \eta(dt_p), \qquad k = 1, \ldots, n,$$

and in particular,

$$\|\eta(T)\|^2 = \sum_{k=1}^{n} \|\eta(S_k)\|^2 = \sum_{k=1}^{n} p! \int \cdots \int_{S_k \times \cdots \times S_k} |\varphi(t_1, \ldots, t_p)|^2 \, dt_1 \cdots dt_p.$$

The final equality shows that $\eta(T) \neq 0$ only for $p \leq 1$, since for $p > 1$ the right-hand side can be made arbitrarily small by a suitable choice of domains S_k of small diameter so that the union $\bigcup_{k=1}^{n} (S_k \times \cdots \times S_k)$ is contained in a sufficiently small neighborhood of the "diagonal" $\{t_1 = \cdots = t_p\}$ of the product $T \times \cdots \times T$ (p times).

Thus there exist only linear additive functionals on white noise, i.e., those for which

$$\xi(S) = \int_S \varphi(t) \eta(dt) + E\xi(S), \qquad S \subseteq T, \tag{1.41}$$

with the real-valued function $\varphi(t)$, $t \in T$, square integrable. Here $\xi(S)$ is a Gaussian variable and a simple calculation gives

$$E \exp[\xi(S)] = 1,$$

while

$$E\xi(S) = -\frac{1}{2} \int_S |\varphi(t)|^2 \, dt.$$

It is also easy to check that the distribution Q of the form (1.35) with density $p = \exp[\xi(T)]$ with respect to the white noise distribution P is Gaussian and differs from P only in mean value, namely

$$\int_\Omega (u, \dot{\eta}) Q(d\omega) = \int_T u(t)\varphi(t) \, dt. \tag{1.42}$$

From the point of view of the Markov property, such a transformation of the distribution is trivial, so that starting from white noise we cannot show the existence of new Markov fields by means of the transformation (1.35).

Later on we show various classes of continuous random functions $\xi(t)$, $t \in T$, in one phase space E or another, having the Markov property—more precisely, the σ-algebras $\mathscr{A}(S)$, $S \subseteq T$, generated by the variables $\xi(t)$, $t \in S$, form a Markov field. For these functions there is a fairly rich supply of additive functionals; a functional of the type

$$\xi(S) = \int_S \varphi(t, \xi(t)) \mu(dt), \qquad S \subseteq T,$$

serves as an example of these, where φ is a real function on the product $T \times E$ and μ is a measure on the space T.

§2. Stopping σ-algebras. Random Sets and the Strong Markov Property

1. Stopping σ-algebras

Let $\mathscr{A}(x)$, $x \in X$, be σ-algebras of events in a probability space $(\Omega, \mathscr{A}, P)$ depending on parameters $x \in X$; we will assume that X does not consist of more than a countable number of elements x. Let $\xi = \xi(\omega)$, $\omega \in \Omega$, be a stopping random variable with values in X, i.e., a mapping of Ω into X such that

$$\{\xi = x\} = \{\omega \colon \xi(\omega) = x\} \in \mathscr{A}, \qquad x \in X.$$

A *stopping σ-algebra* $\mathscr{A}(\xi)$ is defined to be the collection of all events of the form

$$A = \bigcup_x A(x) \cdot \{\xi = x\}, \qquad A(x) \in \mathscr{A}(x), x \in X. \tag{2.1}$$

It is evident that complements and countable unions of events of the form (2.1) are also of this form; for instance, for the event A above the complement will be

$$A^c = \bigcup_x A^c(x) \cdot \{\xi = x\}.$$

Clearly, the stopping σ-algebra $\mathscr{A}(\xi)$ consists of all events A for which

$$A \cdot \{\xi = x\} = A(x) \cdot \{\xi = x\}, \qquad A(x) \in \mathscr{A}(x),$$

for each $x \in X$. This characterization of events $A \in \mathscr{A}(\xi)$ is simplified if the variable $\xi \in X$ is *compatible* with the σ-algebras $\mathscr{A}(x)$, $x \in X$, in the sense that

$$\{\xi = x\} \in \mathscr{A}(x), \qquad x \in X; \tag{2.2}$$

in this case the stopping σ-algebra $\mathscr{A}(\xi)$ consists of all events A such that

$$A \cdot \{\xi = x\} \in \mathscr{A}(x). \tag{2.3}$$

Indeed,

$$A = A \cdot \left[\bigcup_x \{\xi = x\} \right] = \bigcup_x [A \cdot \{\xi = x\}] = \bigcup_x [A(x) \cdot \{\xi = x\}],$$

where $A(x) = A \cdot \{\xi = x\} \in \mathscr{A}(x)$ by the compatibility condition (2.2).
Let

$$\{\mathscr{A}_1(x), \mathscr{B}(x), \mathscr{A}_2(x)\}, \qquad x \in X,$$

be σ-algebras of events such that for each fixed x the corresponding triple has the Markov property—in other words, $\mathscr{B}(x)$ splits $\mathscr{A}_1(x)$ and $\mathscr{A}_2(x)$; we will assume that

$$\mathscr{B}(x) \subseteq \mathscr{A}_1(x), \mathscr{A}_2(x),$$

(see (1.3)). Let the variable $\xi \in X$ be compatible with the σ-algebras $\mathscr{A}_1(x)$, $x \in X$. Then the following proposition is true.

Lemma 1. *The stopping σ-algebras*

$$\{\mathscr{A}_1(\xi), \mathscr{B}(\xi), \mathscr{A}_2(\xi)\}$$

have the Markov property.

PROOF. We must show that our σ-algebras $\{\mathscr{A}_1, \mathscr{B}, \mathscr{A}_2\}$ satisfy equation (1.1) or, equivalently, satisfy (1.4). Using the compatibility condition (2.2) we will prove (1.4) holds.
   Denote by $\tilde{\mathscr{B}}(x)$ the σ-algebra of events generated by the σ-algebra $\mathscr{B}(x)$ and the event $\{\xi = x\}$; it will consist of all events of the form

$$\tilde{B} = B \cdot \{\xi = x\} \cup B' \cdot \{\xi \neq x\}; \qquad B, B' \in \mathscr{B}(x);$$

clearly,

$$C = \tilde{B} \cdot \{\xi = x\} = B(x) \cdot \{\xi = x\}, \qquad B(x) \in \mathscr{B}(x),$$

for every $\tilde{B} \in \tilde{\mathscr{B}}(x)$. Similarly, for every $B \in \mathscr{B}(\xi)$ we have

$$C = B \cdot \{\xi = x\} = B(x) \cdot \{\xi = x\}, \qquad B(x) \in \mathscr{B}(x).$$

From this we see that

$$\tilde{\mathscr{B}}(x) \cdot \{\xi = x\} = \mathscr{B}(\xi) \cdot \{\xi = x\},$$

the collections of events in $\tilde{\mathscr{B}}(x)$ and $\mathscr{B}(\xi)$ which are subsets of the event $\{\xi = x\}$ coincide. Thus, as a function of $\omega \in \{\xi = x\}$, the conditional probability $P\{A \mid \tilde{\mathscr{B}}(x)\}$ is measurable with respect to $\mathscr{B}(\xi)$ and conversely, $P\{A \mid \mathscr{B}(\xi)\}$ is measurable with respect to $\tilde{\mathscr{B}}(x)$ on the set $\{\omega: \xi = x\}$. From this we have

$$\int_C P\{A \mid \tilde{\mathscr{B}}(x)\} P(d\omega) = P(AC) = \int_C P\{A \mid \mathscr{B}(\xi)\} P(d\omega)$$

for any event $C \subseteq \{\xi = x\}$ in $\tilde{\mathscr{B}}(x)$ and $\mathscr{B}(\xi)$, and we conclude that

$$P\{A \mid \tilde{\mathscr{B}}(x)\} = P\{A \mid \mathscr{B}(\xi)\}$$

almost everywhere on the set $\{\omega: \xi = x\}$. Notice that if $A \subseteq \{\xi = x\}$, then each of the indicated conditional probabilities is equal to 0 outside the set $\{\xi = x\}$.

In the same way as for $\tilde{\mathscr{B}}(x)$, we introduce extended σ-algebras of events $\tilde{\mathscr{A}}_1(x)$ and $\tilde{\mathscr{A}}_2(x)$, obtained by adjoining the event $\{\xi = x\}$ to $\mathscr{A}_1(x)$ and $\mathscr{A}_2(x)$. In what follows we will need the Markov property for the triple $\{\tilde{\mathscr{A}}_1(x), \tilde{\mathscr{B}}(x), \tilde{\mathscr{A}}_2(x)\}$; but we know (cf. (1.7)) that for this extension of the Markov triple $\{\mathscr{A}_1(x), \mathscr{B}(x), \mathscr{A}_2(x)\}$, the Markov property is preserved if the event $\{\xi = x\}$ belongs to $\mathscr{A}_1(x)$ or $\mathscr{A}_2(x)$ and this is true by the compatibility condition (2.2):

$$\tilde{\mathscr{A}}_1(x) = \mathscr{A}_1(x), \qquad x \in X.$$

Thus for all $x \in X$

$$P\{A \mid \tilde{\mathscr{A}}_1(x)\} = P\{A \mid \tilde{\mathscr{B}}(x)\}$$

for $A \in \tilde{\mathscr{A}}_2(x)$. We will first take

$$A = A_2 = A_2(x) \cdot \{\xi = x\}.$$

Taking into account that the event $\{\xi = x\}$ belongs to all the stopping σ-algebras and that for an event $A_1 \in \mathscr{A}_1(\xi)$ we have

$$A_1 \cdot \{\xi = x\} = A_1(x) \cdot \{\xi = x\} \in \tilde{\mathscr{A}}_1(x),$$

we get the following chain of equalities:

$$\int_{A_1} P\{A_2 \mid \mathscr{A}_1(\xi)\} P(d\omega) = P(A_1 \cdot A_2) = P(A_1 \cdot \{\xi = x\} \cdot A_2)$$

$$= \int_{A_1 \cdot \{\xi = x\}} P\{A_2 \mid \tilde{\mathscr{A}}_1(x)\} P(d\omega)$$

$$= \int_{A_1 \cdot \{\xi = x\}} P\{A_2 \mid \tilde{\mathscr{B}}(x)\} P(d\omega).$$

As was shown earlier, for the chosen event $A_2 \subseteq \{\xi = x\}$, the conditional probability with respect to the σ-algebra $\tilde{\mathscr{B}}(x)$ coincides with the conditional probability with respect to $\mathscr{B}(\xi)$ and our sequence of equalities can be continued:

$$\int_{A_1 \cdot \{\xi = x\}} P\{A_2 | \tilde{\mathscr{B}}(x)\} P(d\omega) = \int_{A_1 \cdot \{\xi = x\}} P\{A_2 | \mathscr{B}(\xi)\} P(d\omega)$$

$$= \int_{A_1} P\{A_2 | \mathscr{B}(\xi)\} P(d\omega).$$

In short, we have

$$P\{A_2 | \mathscr{A}_1(\xi)\} = P\{A_2 | \mathscr{B}(\xi)\}$$

for $A_2 = A_2 \cdot \{\xi = x\}$. This equality can be extended to all events $A_2 \in \mathscr{A}_2(\xi)$ given by the general formula (2.1), $A_2 = \bigcup_x A_2(x) \cdot \{\xi = x\}$, because for the disjoint events on the right-hand side the sum of the conditional probabilities gives us the conditional probability of the event A_2. This completes the proof of the lemma.

Now we define stopping σ-algebras of events from collections $\{\mathscr{A}(x), x \in X_n\}, n = 1, 2, \ldots$, for a sequence of variables $\xi = \xi_n, n = 1, 2, \ldots$, of the form

$$\xi_n = \varphi_n(\xi_{n+1}), \tag{2.4}$$

where, as before, each variable $\xi_n \in X_n$ takes values in some space X_n having no more than a countable number of elements $x \in X_n$; moreover $\varphi_n(x)$, $x \in X_{n+1}$, is a single-valued mapping $X_{n+1} \rightarrow X_n$ such that

$$\mathscr{A}(x) \subseteq \mathscr{A}(\varphi_n(x)), \qquad x \in X_{n+1}, n = 1, 2, ,\ldots \tag{2.5}$$

In our previous definition of stopping σ-algebra $\mathscr{A}(\xi)$ for a random variable ξ (cf. (2.1)) ξ turned out to be measurable with respect to $\mathscr{A}(\xi)$. Let $\mathscr{B}$ denote the minimal σ-algebra with respect to which all the variables ξ_n, $n = 1, 2, \ldots$, are measurable and consider the union $\mathscr{A}(\xi_n) \vee \mathscr{B}$.

We notice immediately that under the compatibility condition (2.2) for $\xi = \xi_n$ we have

$$\mathscr{B} \subseteq \mathscr{A}(\xi_n), \qquad n = 1, 2, \ldots, \tag{2.6}$$

so that $\mathscr{A}(\xi_n) \vee \mathscr{B} = \mathscr{A}(\xi_n)$. In fact, from (2.4) and (2.5) we have, for any $m > n$, and for the composition φ_{nm} of the mappings $\varphi_k, k = n, \ldots, m - 1$,

$$\xi_n = \varphi_{nm}(\xi_m),$$

$$\mathscr{A}(x) \subseteq \mathscr{A}(\varphi_{nm}(x)), \qquad x \in X_m;$$

the general definition of a sampling σ-algebra (2.1) implies that

$$\{\xi_m = x\} = \{\xi_m = x\} \cdot \{\xi_n = \varphi_{nm}(x)\} \in \mathscr{A}(\xi_n),$$

since from the compatibility condition we have

$$(\xi_m = x\} \in \mathcal{A}(x) \subseteq \mathcal{A}(\varphi_{nm}(x)).$$

Finally we note that $\xi_k = \varphi_{kn}(\xi_n)$ for $k < n$.

Let us look now at general σ-algebras of the type $\mathcal{A}(\xi_n) \vee \mathcal{B}$, not requiring compatibility in them. We will show that they are monotonically decreasing:

$$\mathcal{A}(\xi_{n+1}) \vee \mathcal{B} \subseteq \mathcal{A}(\xi_n) \vee \mathcal{B}, \qquad n = 1, 2, \ldots.$$

By (2.1) it is sufficient that check for $A \in \mathcal{A}(x)$

$$A \cdot \{\xi_{n+1} = x\} = A \cdot \{\xi_{n+1} = x\} \cdot \{\xi_n = \varphi_n(x)\}$$
$$= [A \cdot \{\xi_n = \varphi_n(x)\}] \cdot \{\xi_{n+1} = x\} \in \mathcal{A}(\xi_n) \vee \mathcal{B},$$

but this is so because $A \in \mathcal{A}(x) \subseteq \mathcal{A}(\varphi_n(x))$, $A \cdot \{\xi_n = \varphi_n(x)\} \in \mathcal{A}(\xi_n)$ and $\{\xi_{n+1} = x\} \in \mathcal{B}$.

We define the *stopping σ-algebra* $\mathcal{A}(\xi)$ as the intersection

$$\mathcal{A}(\xi) = \bigcap_n [\mathcal{A}(\xi_n) \vee \mathcal{B}]. \tag{2.7}$$

If compatibility (2.2) holds for $\xi = \xi_n$, $n = 1, 2, \ldots$, we have $\mathcal{A}(\xi_n) \vee \mathcal{B} = \mathcal{A}(\xi_n)$ and our stopping σ-algebra $\mathcal{A}(\xi) = \bigcap_n \mathcal{A}(\xi_n)$ consists of all those events A such that (cf. (2.3))

$$A \cdot \{\xi_n = x\} \in \mathcal{A}(x), \qquad x \in X_n, n = 1, 2, \ldots.$$

We will consider the question of the Markov behavior of the stopping σ-algebras

$$\{\mathcal{A}_1(\xi), \mathcal{B}(\xi), \mathcal{A}_2(\xi)\} \tag{2.8}$$

from the Markov collections

$$\{\mathcal{A}_1(x), \mathcal{B}(x), \mathcal{A}_2(x)\}, \qquad x \in X_n, n = 1, 2, \ldots,$$

assuming condition (2.5) for

$$\mathcal{A}(x) = \mathcal{A}_1(x), \mathcal{B}(x), \mathcal{A}_2(x)$$

as well as (2.2), the compatibility of the sequence $\xi = \xi_n$ with the σ-algebras $\mathcal{A}(x) = \mathcal{A}_1(x)$, $x \in X_n$, $n = 1, 2, \ldots$.

As we saw already (cf. (2.6)) $\mathcal{B} \subseteq \mathcal{A}_1(\xi)$, the σ-algebra $\mathcal{B}(\xi_n) \vee \mathcal{B}$ splits $\mathcal{A}_1(\xi) \subseteq \mathcal{A}_1(\xi_n)$ and $\mathcal{A}_2(\xi) \subseteq \mathcal{A}_2(\xi_n) \vee \mathcal{B}$. Thus these properties hold for the intersection

$$\mathcal{B}(\xi) = \bigcap_n [\mathcal{B}(\xi_n) \vee \mathcal{B}]$$

of the monotonically decreasing σ-algebras $\mathcal{B}(\xi_n) \vee \mathcal{B}$ and we have the following result;

Lemma 2. *The stopping σ-algebras (2.8) have the Markov property.*

2. Random Sets

Let T be a separable, locally compact metric space. By a *random set* $\xi \subseteq T$ we mean a mapping $\xi = \xi(\omega)$, $\omega \in \Omega$, of our probability space Ω into sets $S \subseteq T$ such that for all open domains $S \subseteq T$ the events

$$\{\xi \subseteq S\} = \{\omega : \xi(\omega) \subseteq S\}$$

are defined. Let $\mathscr{A}$ be some σ-algebra containing all these events; we will say that the random set is *measurable* with respect to $\mathscr{A}$.

We will consider only random sets for which, with probability 1, $\xi(\omega)$ is relatively compact or has compact complement.

Let ξ be a random set in the space T and for any open domain $S \subseteq T$

$$\{\xi \subseteq S\} \in \mathscr{A}. \tag{2.9}$$

For a closed set $S \subseteq T$ the inclusion $\xi(\omega) \subseteq S$ is equivalent to $\xi(\omega) \subseteq S^{\varepsilon_n}$, $\varepsilon_n \to 0$, and so

$$\{\xi \subseteq S\} = \bigcap_n \{\xi \subseteq S^{\varepsilon_n}\} \in \mathscr{A};$$

here, as usual, S^{ε} denotes an ε-neighborhood of the set S.

The closure $\bar{\xi}$ is also a random set since the inclusion $\bar{\xi}(\omega) \subseteq S$ for open $S = \bigcup_n \overline{S^{-\varepsilon_n}}$ is equivalent to $\xi(\omega) \subseteq \overline{S^{-\varepsilon_n}}$ for at least one $n = 1, 2, \ldots$, where $S^{-\varepsilon}$ denotes the set of all those points $t \in S$ whose distance to the boundary of the domain S is greater than ε, and

$$\{\bar{\xi} \subseteq S\} = \bigcup_n \{\xi \subseteq \overline{S^{-\varepsilon_n}}\} \in \mathscr{A}. \tag{2.10}$$

We note that if the random set ξ is closed (with probability 1) then according to (2.10) it will be measurable with respect to the σ-algebra of events $\mathscr{A}$ if the inclusion (2.9) is true for closed $S \subseteq T$.

It is easy to see that the events defined by

$$\{\xi \cap S = \varnothing\} = \{\xi \subseteq S^c\} \in \mathscr{A},$$
$$\{\xi \cap S \neq \varnothing\} = \{\xi \not\subseteq S^c\} = \{\xi \subseteq S^c\}^c \in \mathscr{A}, \tag{2.11}$$

are in $\mathscr{A}$ for open or closed $S \subseteq T$. In fact, if both of the conditions in (2.11) hold for all closed S, then ξ is measurable with respect to the σ-algebra $\mathscr{A}$. In the case where ξ is closed, the preceding statement holds with "closed" changed to "open"—see the remark following (2.10); moreover, this can be restricted to sets $S = U_n$, $n = 1, 2, \ldots$, forming a complete neighborhood basis, since

$$\{\xi \cap S \neq \varnothing\} = \bigcup_{n:\, U_n \subseteq S} \{\xi \cap U_n \neq \varnothing\},$$

i.e., if $\xi \cap S \neq \varnothing$ then there is some $t \in \xi \cap S$ and neighborhood $U_n \subseteq S$ containing the point t.

Clearly, the measurability of a closed ξ with respect to the σ-algebra of events $\mathcal{A}$ is equivalent to the measurability with respect to $\mathcal{A}$ of the random variables $\rho(t, \xi), t \in T$, each of which is the distance from a fixed point t to the random set ξ; this is because

$$\{\rho(t, \xi) < r\} = \{\xi \cap t^r \neq \varnothing\}, \qquad r > 0, \qquad (2.12)$$

where t^r is the r-neighborhood of the point t.

In the case of a closed random domain† ξ, its measurability is equivalent to the measurability of the random indicator function $1_\xi(t), t \in T$:

$$1_\xi(t) = \begin{cases} 1, & \text{if } t \in \xi, \\ 0, & \text{if } t \notin \xi, \end{cases} \qquad (2.13)$$

since the event $t \in \xi$ denotes the nonempty intersection of the set ξ with the singleton set $S = \{t\}$; for a domain ξ and any open $S \subseteq T$, a nonempty intersection $\xi \cap S$ contains some entire neighborhood, so that taking a countable dense set of points t_n in S we have

$$\{\xi \cap S \neq \varnothing\} = \bigcup_n \{1_\xi(t_n) = 1\} \in \mathcal{A}. \qquad (2.13a)$$

The analogous statement is also true for open domains ξ since then (2.13a) is valid for closed S and a countable dense sets of points t_n in S.

In the case of a closed ξ we have, for any set $S \subseteq T$,

$$\{\xi \supseteq S\} \in \mathcal{A}; \qquad (2.14)$$

indeed, the inclusion $\xi(\omega) \supseteq S$ means that $\xi(\omega) \cap U_n \neq \varnothing$ for all U_n in a neighborhood basis for the space T for which $U_n \cap S \neq \varnothing$—i.e., if in S there is a point $t \notin \xi(\omega)$, then there is a neighborhood U_n containing t with $U_n \cap \xi(\omega) = \varnothing$, consequently

$$\{\xi \supseteq S\} = \bigcap_n \{\xi \cap U_n \neq \varnothing\} \in \mathcal{A}.$$

The complement ξ^c of a closed random set ξ is measurable:

$$\{\xi^c \subseteq S\} = \{\xi \supseteq S^c\} \in \mathcal{A} \qquad (2.15)$$

for any $S \subseteq T$. From this it follows, in particular, that for the measurability of an open random set ξ with respect to the σ-algebra of events $\mathcal{A}$ the inclusion

$$\{\xi \supseteq S\} = \{\xi^c \subseteq S^c\} \in \mathcal{A} \qquad (2.16)$$

for open (or closed) sets S is sufficient; $\xi = \eta^c$ is the complement of the closed $\eta = \xi^c$, which is measurable with respect to $\mathcal{A}$.

Now let the random set ξ be such that with probability 1 its boundary $\partial \xi$ coincides with the boundary $\partial \bar{\xi}$ of the closure $\bar{\xi}$:

$$\partial \xi = \partial \bar{\xi}. \qquad (2.17)$$

† Recall that by a *domain* we mean a set which is contained in the closure of its interior.

This means that the interior domains $\overset{\circ}{\xi}$ and $\overset{\circ}{\bar{\xi}}$ coincide, since if a point s belongs to the boundary $\partial\xi$ but is in the interior of $\bar{\xi}$ then $\overset{\circ}{\xi} \neq \overset{\circ}{\bar{\xi}}$ and conversely.

In the case when the set ξ is a domain, $\xi = \overset{\circ}{\bar{\xi}}$, (2.17) gives the following equalities:

$$\partial\overset{\circ}{\xi} = \partial\overset{\circ}{\xi^c} = \partial\bar{\xi} = \partial\bar{\xi}^c, \tag{2.18}$$

where the (open) complement $\eta = \overset{\circ}{\xi^c}$ of the closure $\bar{\xi}$ has its closure $\bar{\eta} = \overset{\circ}{\bar{\xi}^c}$; clearly, the closure $\bar{\eta}$ of the measurable complement $\eta = \overset{\circ}{\xi^c}$ is measurable with respect to $\mathscr{A}$ and so is the interior domain $\overset{\circ}{\xi} = \bar{\eta}^c$. From this, for every open domain S we have

$$\{\xi \supseteq S\} = \{\overset{\circ}{\xi} \supseteq S\} = \{\bar{\xi} \supseteq S\} \in \mathscr{A}, \tag{2.19}$$

since by the inclusion $\bar{\xi} \supseteq S$ each point $t \in S$ is an interior point in $\overset{\circ}{\bar{\xi}} = \overset{\circ}{\xi}$ and $\overset{\circ}{\xi} \supseteq S, \bar{\xi} \supseteq S$. In particular, (2.17) is always satisfied by a closed domain ξ and thus a closed domain ξ will be measurable with respect to a σ-algebra of events $\mathscr{A}$ if inclusion (2.19) is true for any open domain $S \subseteq T$: the interior domain $\overset{\circ}{\xi}$ is measurable (cf. (2.16)) and $\xi = \overset{\circ}{\bar{\xi}}$.

The intersection $\eta = \xi \cap S$ of a random set ξ with a fixed closed set $S \subseteq T$ is measurable with respect to the same σ-algebra $\mathscr{A}$ as ξ is, since

$$\{\eta \cap S' = \varnothing\} = \{\xi \cap (S \cap S') = \varnothing\}$$

for any closed $S' \subseteq T$—see (2.11).

Let random sets ξ_1, ξ_2 be measurable with respect to $\mathscr{A}_1, \mathscr{A}_2$; then the union $\xi_1 \cup \xi_2$ is measurable with respect to the σ-algebra $\mathscr{A}_1 \vee \mathscr{A}_2$ because

$$\{\xi_1 \cup \xi_2 \subseteq S\} = \{\xi_1 \subseteq S\} \cdot \{\xi_2 \subseteq S\}$$

for any $S \subseteq T$. The intersection $\xi_1 \cap \xi_2$ is measurable with respect to $\mathscr{A}_1 \vee \mathscr{A}_2$ when ξ_1, ξ_2 are open or closed domains because, by (2.13),

$$1_{\xi_1 \cap \xi_2}(t) = 1_{\xi_1}(t) \cdot 1_{\xi_2}(t)$$

for any $t \in T$. When ξ_1 and ξ_2 are closed sets with at least one of them (say ξ_1, for convenience) compact, then the intersection $\xi_1 \cap \xi_2$ is also measurable with respect to the σ-algebra of events $\mathscr{A}_1 \vee \mathscr{A}_2$ since for every closed $S \subseteq T$ the intersections $\eta_1 = \xi_1 \cap S, \eta_2 = \xi_2 \cap S$ are measurable with respect to $\mathscr{A}_1$ and $\mathscr{A}_2$ and

$$\{(\xi_1 \cap \xi_2) \cap S = \varnothing\} = \{\eta_1 \cap \eta_2 = \varnothing\} = \bigcup_n [\{\eta_1 \subseteq S_n\} \cap \{\eta_2 \subseteq S_n^c\}],$$

where $S_n, n = 1, 2, \ldots$, are all possible finite unions of members of a complete neighborhood basis in T (cf. (2.11)).

In particular, it follows from condition (2.17) that the set $\partial\xi = \bar{\xi} \cap \bar{\eta}$, together with $\eta = \overset{\circ}{\xi^c}$, is measurable with respect to the σ-algebra $\mathscr{A}$ because one of the sets $\bar{\xi}, \bar{\eta}$ is compact by our earlier assumption.

We remark that in general the boundary $\partial\xi$, the interior domain $\overset{\circ}{\xi}$ and events of the form $\{\xi \supseteq S\}$ are not measurable if condition (2.17) is not satisfied. For an example of this we can take the open domain $\xi = T\backslash\tau$, where τ is some random point in T such that $\{\tau = t\}$ is an event of probability 0 for each fixed $t \in T$. In this situation, all events $\{\xi \subseteq S\}$ have probability 0 if $S \neq T$ and so the random domain ξ is measurable with respect to the trivial σ-algebra, but this cannot be said about its boundary $\partial\xi = \tau$.

Let ξ be a compact random domain which is measurable with respect to a σ-algebra of events $\mathscr{A}$ and let ξ_0 be its connected component containing a fixed point $t_0 \in T$. Obviously, ξ_0 is also a compact domain.

Lemma 3. *The connected component ξ_0 is measurable with respect to $\mathscr{A}$.*

PROOF. We take a complete countable neighborhood basis for the space T. We will say that the point t *is connected by an ε-chain to* t_0 through the set S if we can find a finite number of basic neighborhoods $U_1, \ldots, U_n$ of diameter $\leq \varepsilon$ and having the following property: $t_0 \in U_1$, each succeeding U_k has a nonempty intersection with U_{k-1}, $t \in U_n$ and in addition each U_k has a nonempty intersection with S. The compact connected domain ξ_0 consists of all points $t \in \xi$ such that for any $\varepsilon > 0$ there is an ε-chain through ξ connecting t with t_0. For fixed t the existence of an ε-chain through ξ is an event A_ε in the σ-algebra $\mathscr{A}$. In fact, there are altogether only a countable number of ε-chains through T and it turns out that the ε-chain through ξ is the nth "ε-chain through T" for some n and is an event $B_\varepsilon^{(n)}$ in $\mathscr{A}$, so that $A_\varepsilon = \bigcup_n B_\varepsilon^{(n)} \in \mathscr{A}$. By taking a sequence $\varepsilon = \varepsilon_n \to 0$ we get that a fixed point t being an element of the connected component ξ_0 is an event $A = \bigcap_\varepsilon A_\varepsilon \in \mathscr{A}$, that is, the random indicator functions $1_{\xi_0}(t)$, $t \in T$, are measurable with respect to $\mathscr{A}$, together with the closed random domain ξ_0 (cf. (2.13)). $\qquad\square$

3. Compatible Random Sets

We suppose that each open domain $S \subseteq T$ in a separable locally compact metric space T is associated with a σ-algebra of events $\mathscr{A}(S)$ and

$$\mathscr{A}(S_1) \subseteq \mathscr{A}(S_2) \quad \text{if } S_1 \subseteq S_2. \tag{2.20}$$

We consider random sets $\xi \subseteq T$ which are *compatible* with this family of σ-algebras $\mathscr{A}(S)$, $S \subseteq T$, in the sense that

$$\{\xi \subseteq S\} \in \mathscr{A}(S^\varepsilon), \qquad \varepsilon > 0, \tag{2.21}$$

for all open domains S.

The compatibility condition (2.21) can be extended to closed sets S; what is more, a closed set ξ is compatible if (2.21) is satisfied for closed S, as we basically saw in (2.10). Furthermore, if ξ is compatible then so is the closure $\bar{\xi}$ and, by (2.17), the interior domain $\overset{\circ}{\xi}$ since for every open $S \subseteq T$

$$\{\bar{\xi} \subseteq \bar{S}\} = \{\overset{\circ}{\bar{\xi}} \subseteq S\} = \{\overset{\circ}{\xi} \subseteq S\}.$$

For every open or closed $S' \subseteq S$ we have

$$\{\xi \subseteq S, \xi \not\subseteq S'\} = \{\xi \subseteq S\}\backslash\{\xi \subseteq S'\} \in \mathcal{A}(S^\varepsilon), \qquad \varepsilon > 0. \qquad (2.22)$$

In addition to (2.22) we note that for closed ξ, closed S and any set $S' \subseteq S$ we have

$$\{\xi \subseteq S, \xi \supseteq S'\} = \bigcap_n \{\xi \subseteq S, \xi \cap U_n \neq \varnothing\} \in \mathcal{A}(S^\varepsilon), \qquad \varepsilon > 0,$$

where the intersection is taken over all U_n in a complete neighborhood basis for which $U_n \cap S' \neq \varnothing$, and

$$\{\xi \subseteq S, \xi \cap U_n \neq \varnothing\} = \{\xi \subseteq S\}\backslash\{\xi \subseteq S\backslash U_n\} \in \mathcal{A}(S^\varepsilon), \qquad \varepsilon > 0.$$

A random set $\xi \subseteq T$ will be called *co-compatible* with the family of σ-algebras $\mathcal{A}(S)$, $S \subseteq T$, if

$$\{\xi \supseteq S\} \in \mathcal{A}(S^\varepsilon), \qquad \varepsilon > 0, \qquad (2.23)$$

for every open domain S; in the case of closed ξ the condition of co-compatibility automatically carries over to every domain S because

$$\{\xi \supseteq S\} = \{\xi \supseteq \bar{S}\} = \{\xi \supseteq \mathring{S}\},$$

where $\mathring{S}$ is the interior domain of S.

Obviously, the interior domain of a co-compatible set ξ is also co-compatible; for open S, $\{\xi \supseteq S\} = \{\mathring{\xi} \supseteq S\}$.

We introduce some examples of co-compatible random sets which are important for applications.

Let $\xi(t)$, $t \in T$, be a random function with values in some topological ("phase") space and $\mathcal{A}(S)$ be the σ-algebra of events generated by the trajectory of $\xi(t)$, $t \in S$, S a domain in T. We suppose that for each $t \in T$, we are interested in some "local" property A_t, defined by values of our random function in an infinitesimal neighborhood of the point t, such as continuity or whether values of $\xi(t)$ lie in some specific set in the phase space. Let $\xi \subseteq T$ be the set of all those points for which the trajectory $\xi(t)$, $t \in T$, has this particular "local" property. If we ignore the question of measurability of the image $\xi = \xi(\omega)$, $\omega \in \Omega$, then it is clear that the random set ξ is co-compatible —the inclusion $S \subseteq \xi$ means that the trajectory has the desired property A_t for all points $t \in S$ and this is an event in the σ-algebra $\mathcal{A}(S^\varepsilon)$, $\varepsilon > 0$. To state this more precisely, we assume that the set ξ is a closed domain with probability 1 and the presence of the local property under consideration at a point $t \in S$ is an event A_t in the σ-algebra $\mathcal{A}(S^\varepsilon)$, $\varepsilon > 0$. Then ξ is a co-compatible random set because the inclusion $\{\xi \supseteq S\}$ is equivalent to the inclusion $\{\xi \supseteq S'\}$, where $S' \subseteq S$ is a countable dense set in S and the event

$$\{\xi \supseteq S'\} = \bigcup_{t \in S'} A_t \in \mathcal{A}(S^\varepsilon)$$

is the union of a countable number of events $A_t \in \mathcal{A}(S^\varepsilon)$; in the case of a closed domain ξ, condition (2.23) gives measurability with respect to the σ-algebra of events $\mathcal{A}(T)$ (see (2.19)).

An important class of compatible random sets is described below.

Theorem 1. *Let ξ be a compact random domain, co-compatible with the family $\mathscr{A}(S)$, $S \subseteq T$. Then the connected component ξ_0 containing a fixed point $t_0 \in T$ is compatible with this family.*

PROOF. We take an open domain $x \subseteq T$ and let $\eta = \overset{\circ}{\xi} \cap x$. The open random set η is measurable with respect to $\mathscr{A}(x^\varepsilon)$, $\varepsilon > 0$, so that for any open S the event $\{\eta \supseteq S\}$ is empty if $S \not\subseteq x$ and coincides with the event $\{\xi \supseteq S\} \in \mathscr{A}(x^\varepsilon)$ when $S \subseteq x$, (cf. (2.16)). Therefore $\bar{\eta} = \overline{\overset{\circ}{\xi} \cap x}$ is also measurable with respect to $\mathscr{A}(x^\varepsilon)$.

We denote by η_0 the connected component of the compact random domain $\bar{\eta}$ containing the fixed point $t_0 \in T$. As we have shown, η_0 is measurable with respect to the same σ-algebra of events as $\bar{\eta}$ is (see the lemma in §2.2). Since we have shown that $\bar{\eta} = \overline{\overset{\circ}{\xi} \cap x}$ is measurable with respect to $\mathscr{A}(x^\varepsilon)$, $\varepsilon > 0$, η_0 is thus measurable with respect to $\mathscr{A}(x^\varepsilon)$, $\varepsilon > 0$. The proof will be finished if we show that for any closed $S \subseteq T$

$$\{\xi_0 \subseteq S\} = \bigcup_{1/n < \varepsilon} \{\eta_0^{(n)} \subseteq S\}, \qquad \varepsilon > 0,$$

where

$$\eta^{(n)} = \overline{\overset{\circ}{\xi} \cap x^{(n)}}, \qquad x^{(n)} = S^{1/n}.$$

We know that the components $\eta_0^{(n)}$ are measurable with respect to $\mathscr{A}(S^\varepsilon)$. Since $\eta^{(n)} \subseteq \xi$, $\eta_0^{(n)} \subseteq \xi_0$ always and $\eta_0^{(n)} \subseteq S$ if $\xi_0 \subseteq S$. On the other hand, let $\eta_0^{(n)} \subseteq S$ for some sufficiently large n and suppose that $\xi_0 \not\subseteq S$. Then the domain $\xi_0 \backslash \eta_0^{(n)}$ is not empty and has a limit point $s \in \eta_0^{(n)}$ because otherwise the connected domain ξ_0 would split into two disjoint closed domains $\eta_0^{(n)}$ and $\xi_0 \backslash \eta_0^{(n)}$. Clearly, the point $s \in \eta_0^{(n)} \cap S$ is a limit of points

$$t \in \overset{\circ}{\xi_0} \cap x^{(n)} \subseteq \overline{\overset{\circ}{\xi} \cap x^{(n)}} = \eta^{(n)}, \qquad t \notin \eta_0^{(n)}$$

(as usual, $\overset{\circ}{\xi_0}$ is the interior domain of ξ_0); but this cannot be true for $\eta^{(n)}$ with connected component $\eta_0^{(n)}$, hence the closed set $\eta^{(n)} \backslash \eta_0^{(n)}$ is some positive distance from $\eta_0^{(n)}$. Thus the assumption that $\xi_0 \not\subseteq S$ is false and $\xi_0 \subseteq S$, which is what we wanted to show. □

To conclude, we introduce an example of a compatible random set which is useful in applications. Let $\xi(t)$, $t \in T$, be a continuous random function on a compact T with values in a topological "phase" space, with an initial value $\xi(t_0) = a$ at a fixed point t_0. Suppose that $\xi \subseteq T$ is the set of all points in T for which the trajectory $\xi(t)$, $t \in T$, lies in a fixed open set A in the phase space which contains the value a. It is apparent that the random set $\xi \subseteq T$ is open. Its closure $\bar{\xi}$ is co-compatible with the family of σ-algebras of events $\mathscr{A}(S)$, $S \subseteq T$, each of which is determined by the behavior of the function $\xi(t)$, $t \in S$, on the open domain S; in fact, the inclusion $\{\bar{\xi} \supseteq S\}$ is uniquely defined by the trajectory $\xi(t)$, $t \in S^\varepsilon$, on an infinitesimal neighborhood S^ε:

$$\{\bar{\xi} \supseteq S\} = \{\xi \supseteq S\} \in \mathscr{A}(S^\varepsilon), \qquad \varepsilon > 0.$$

By our theorem, the connected component $\bar{\xi}_0$ of the compact domain ξ containing the initial point t_0 is compatible.

When the parameter space T is the time interval $[t_0, T]$, the connected component $\bar{\xi}_0$ is the interval $[t_0, \tau]$, and we call the time τ, for $\tau < T$, the *first exit time* from the open set A (alternatively, the *first entrance time* to the complement A^c) for the trajectory $\xi(t)$, $t \geq t_0$, originating at the point $\xi(t_0) = a \in A$. For the random process $\xi(t)$, $t \geq t_0$, the compatible random intervals $[t_0, \tau)$ correspond to what are known as stopping times τ—times which are "independent of the future." For such stopping times τ we have the well-known strong Markov property for the random process $\xi(t)$, $t \geq t_0$; we consider below a generalization of this to the case of an arbitrary parameter t in the metric space T.

4. Strong Markov Property

We will be looking at σ-algebras of events $\mathscr{A}(S)$, $S \subseteq T$, of the same type as in the preceding section—each is connected with an open domain S in a separable locally compact metric space T and condition (2.20) holds.

We will also assume that the family $\mathscr{A}(S)$, $S \subseteq T$, is Markov with respect to all open domains $S \subseteq T$ which are pre-compact or have compact complements, i.e., the σ-algebras

$$\{\mathscr{A}(S)^\varepsilon, \mathscr{A}(\partial S)^\varepsilon, \mathscr{A}(S^c)^\varepsilon\}, \qquad \varepsilon > 0, \tag{2.24}$$

have the Markov property. Here we use the new notation

$$\mathscr{A}(S)^\varepsilon = \mathscr{A}(S^\varepsilon)$$

for the σ-algebra of events $\mathscr{A}(S^\varepsilon)$ connected with the neighborhood S^ε of the set $S \subseteq T$; in particular, $\mathscr{A}(\partial S)^\varepsilon$ denotes the σ-algebra of events for the neighborhood Γ^ε of the boundary $\Gamma = \partial S$ of the domains S and $\bar{S}^c$, and $\mathscr{A}(S)^\varepsilon$ and $\mathscr{A}(S^c)^\varepsilon$ are the σ-algebras of events for the neighborhoods $S^\varepsilon = S \cup \Gamma^\varepsilon$ and $S^c = S^c \cup \Gamma^\varepsilon$. Our goal is to define sampling σ-algebras $\mathscr{A}(\xi)^\varepsilon$ for the random set $\xi \subseteq T$ and for all compatible relatively compact domains $\xi \subseteq T$ to establish the Markov property for the stopping σ-algebras

$$\{\mathscr{A}(\xi)^\varepsilon, \mathscr{A}(\partial \xi)^\varepsilon, \mathscr{A}(\xi^c)^\varepsilon\};$$

this will be the *strong Markov property* for the family $\mathscr{A}(S)$, $S \subseteq T$.

For a random set ξ taking a countable number of values $x = x_1, x_2, \ldots$, such stopping σ-algebras can be defined using the construction in §2.1,

$$\mathscr{A}_1(x) = \mathscr{A}(x)^\varepsilon, \qquad \mathscr{B}(x) = \mathscr{A}(\partial x)^\varepsilon, \qquad \mathscr{A}_2(x) = \mathscr{A}(x^c)^\varepsilon,$$

see (2.1); their Markov character was established by Lemma 1.

In the general case we employ the limiting procedure (2.7) for a suitable collection

$$\{\mathscr{A}_1(x), \mathscr{B}(x), \mathscr{A}_2(x)\}, \qquad x \in X_n, n = 1, 2, \ldots,$$

and define stopping σ-algebras of the form (2.8).

We take a sequence of partitions $T = \bigcup_k x_{kn}$ of the space T into closed domains (cells)

$$x_{1n}, x_{2n}, \ldots \tag{2.25}$$

of diameter $\leq \varepsilon_n \to 0$ so that for a given nth partition ($n = 1, 2, \ldots$) no pair of cells have any interior points in common, and for any cells x_{kn} and x_{jm} from different partitions having a common interior point the inclusion

$$x_{jm} \subseteq x_{kn}$$

holds when $m > n$; in addition, each compact set in T has a finite covering of cells from the nth partition, for each n. For example one can construct such partitions in the following fashion. We begin with a complete neighborhood basis $U_1, U_2, \ldots$, each with compact closure. As a first step we take only those neighborhoods having diameter $\leq \varepsilon_1$—let these be $U_{11}, U_{21}, \ldots$. From them we form

$$x_{11} = \bar{U}_{11}, x_{21} = \overline{U_{21} \backslash \bar{U}_{11}}, \ldots, x_{k1} = \overline{U_{k1} \backslash \left(\bigcup_{j<k} U_{j1} \right)};$$

these are the cells for the first partition and it is obvious that they are compact domains. For the second step we consider sets from the original neighborhood basis of diameter $\leq \varepsilon_2$—call them $U_{12}, U_{22}, \ldots$. For each cell x in the first partition we construct new cells of the form

$$\overline{x \cap U_{12}}, \overline{x \cap U_{22} \backslash \bar{U}_{12}}, \ldots, \overline{x \cap U_{k2} \backslash \left(\bigcup_{j<k} \bar{U}_{j2} \right)}, \ldots;$$

since the compact set x has a finite covering of neighborhoods $U_{21}, U_{22}, \ldots$, we get from this a finite partition of each of the original cells x. We remark here for clarity that the intersection of any domain with an open domain again is a domain; thus our cells, having the form

$$\overline{x \cap U_{k2} \backslash \left(\bigcup_{j<k} U_{j2} \right)} = \overline{x \cap U_{k2} \cap \left[\bigcap_{j<k} \bar{U}_{j2}^c \right]},$$

are (compact) domains. Each compact set has a finite covering from U_1, $U_2, \ldots$ and as a consequence, a finite covering by our cells. Proceeding in this way, we get the desired sequence of partitions.

We denote by X_n the collection of all sets $x \subseteq T$ which are finite unions of cells from the nth partition; there are only a countable number of distinct elements $x \in X_n$.

For a compact domain $\xi \subseteq T$ one can find a set $x \in X_n$ containing ξ. Moreover the structure of elements $x \in X_n$ is such that there is a minimal set $\xi_n \in X_n$ containing the domain ξ, that is $\xi_n \subseteq x$ if $x \supseteq \xi$.

In fact, since all sets $x \in X_n$ are closed, the inclusion $x \supseteq \xi$ for compact domains $\xi = \overset{\circ}{\bar{\xi}}$ is equivalent to the inclusion $x \supseteq \overset{\circ}{\xi}$, where $\overset{\circ}{\xi}$ is the interior (open) domain of ξ. If a cell $x_{kn} \in X_n$ has a nonempty intersection with the open domain $\mathring{\xi}$, then this is also true for the interior domain $\mathring{x}_{kn}$ of this cell; thus $x_{kn} \subseteq x$ for $x \supseteq \xi$, so that in X_n there are no cells other than x_{kn} which contain the intersection $\mathring{x}_{kn} \cap \mathring{\xi}$. From this it is evident that the union of all cells x_{kn} which have nonempty intersection with $\mathring{\xi}$ gives the minimal set in X_n containing ξ, and also $\xi = \bar{\mathring{\xi}}$.

The set $\xi_n \in X_n$ we have introduced can be characterized as the collection of all cells x_{kn} which impinge on the interior domain $\mathring{\xi}$, i.e., $x_{kn} \cap \mathring{\xi} \neq \varnothing$.

We introduce a transformation $\varphi_n(S)$ on relatively compact domains $S \subseteq T$ by defining

$$\varphi_n(S) = S_n$$

as the minimal set in X_n which contains S—this set S_n is the union of all cells $x_{kn} \in X_n$ intersecting the interior domain $\mathring{S}$. We denote by $\partial_n S$ the "thick" boundary of the domain S, consisting of all cells $x_{kn} \subseteq S_n$ impinging on an infinitesimal neighborhood of the boundary $\Gamma = \partial S$, more precisely, having nonempty intersection with the domain $\Gamma^\varepsilon \cap \mathring{S}$ for all $\varepsilon > 0$. Finally, we let

$$S^{cn} = S^c \cup \partial_n S.$$

It is easily seen that

$$S_n = \varphi_n(S_m), \qquad \partial_n S = \varphi_n(\partial_m S) \quad \text{for } n \leq m. \tag{2.26}$$

In fact, $S_n = \varphi_n(S) \subseteq \varphi_n(S_m)$ since $S \subseteq S_m$. On the other hand, if the cell x_{kn} belongs to $\varphi_n(S_m)$ then its interior domain $\mathring{x}_{kn}$ has a nonempty intersection with the interior domain of one of the cells $x_{jm} \subseteq S_m = \varphi_m(S)$; but this cell x_{jm} must be entirely contained in x_{kn}, from which it follows that together with x_{jm}, the cell x_{kn} has a nonempty intersection with the interior domain $\mathring{S}$ and belongs to $\varphi_n(S)$. The proof of the second equality in (2.26) is similar: for each cell $x_{kn} \subseteq \partial_n S$ there is some boundary point $t \in \partial S$ such that x_{kn} has a nonempty intersection with $t^\varepsilon \cap \mathring{S}$ for any $\varepsilon > 0$, and in the finite union $x_{kn} = \bigcup_j x_{jm}$ there is some cell x_{jm} having this same property; in particular x_{kn} intersects $x_{jm} \subseteq \partial_m S$ and thus $\partial_n S \subseteq \varphi_n(\partial_m S)$. In the other direction, if the cell x_{kn} impinges on $\partial_m S \subseteq S_m$ then it entirely contains some cell $x_{jm} \subseteq \partial_m S$, as we noted before, and together with x_{jm} has a nonempty intersection with an infinitesimal neighborhood of the boundary ∂S, that is, $x_{kn} \subseteq \partial_n S$, hence $\varphi_n(\partial_m S) \subseteq \partial_n S$. From the definition of the (closed) set $\partial_n S$ it follows that $\partial S \subseteq \partial_n S$ and it is clear, also, that

$$\partial S_n \subseteq \partial_n S.$$

For each $\varepsilon > 0$

$$S_n \subseteq S^\varepsilon, \qquad \partial_n S \subseteq (\partial S)^\varepsilon$$

for sufficiently large n—namely, such that $\varepsilon_n < \varepsilon$, where ε_n is the diameter of cells in the nth partition—because each cell x_{kn} in S_n, $\partial_n S$ contains some point $t \in S$, ∂S respectively. It is evident that

$$S = \bigcap S_n, \qquad \partial S = \bigcap \partial_n S, \qquad \bar{S^c} = \bigcap S^{cn}.$$

We introduce σ-algebras of the form (2.24):

$$\mathscr{A}_1(x) = \mathscr{A}(x)^\varepsilon, \qquad \mathscr{B}(x) = \mathscr{A}(\partial_n x)^\varepsilon, \qquad \mathscr{A}_2(x) = \mathscr{A}(x^{c_n})^\varepsilon, \qquad x \in X_n;$$

$$\tag{2.27}$$

these constitute our nth collection, $n = 1, 2, \ldots$, in accordance with the notation in §2.1. We consider stopping σ-algebras

$$\mathscr{A}_1(\xi_n) = \mathscr{A}(\xi_n)^\varepsilon, \qquad \mathscr{B}(\xi_n) = \mathscr{A}(\partial_n \xi)^\varepsilon, \qquad \mathscr{A}_2(\xi_n) = \mathscr{A}(\xi^{c_n})^\varepsilon, \qquad \xi_n = \varphi_n(\xi);$$

the general form of these was described by (2.1). It is apparent that conditions (2.4) and (2.5) will be satisfied for the σ-algebras (2.27); this is an immediate consequence of the monotonicity (2.20) and equation (2.26).

We define stopping σ-algebras as in §2.1 using the limiting process (2.7), namely,

$$\mathscr{A}_1(\xi) = \mathscr{A}(\xi)^\varepsilon = \bigcap_n \left[\mathscr{A}(\xi_n)^\varepsilon \vee \mathscr{B} \right],$$

$$\mathscr{B}(\xi) = \mathscr{A}(\partial \xi)^\varepsilon = \bigcap_n \left[\mathscr{A}(\partial_n \xi)^\varepsilon \vee \mathscr{B} \right], \tag{2.28}$$

$$\mathscr{A}_2(\xi) = \mathscr{A}(\xi^c)^\varepsilon = \bigcap_n \left[\mathscr{A}(\xi^{c_n})^\varepsilon \vee \mathscr{B} \right],$$

where $\mathscr{B}$ is the smallest σ-algebra containing all events of the form

$$\{\xi_n = x\}, \qquad x \in X_n, n = 1, 2, \ldots; \tag{2.29}$$

note here that

$$\xi = \bigcap_n \xi_n, \qquad \partial \xi = \bigcap_n \partial_n \xi, \qquad \bar{\xi}^c = \bigcap_n \xi^{c_n}.$$

We intend to make use of Lemma 2, §2.1, in connection with the stopping σ-algebras (2.28), and for this we must have compatibility of the ξ_n with the families $\mathscr{A}(x)^\varepsilon$, $x \in X_n$, $n = 1, 2, \ldots$.

We assume that the random domain ξ is compatible with σ-algebras $\mathscr{A}(S)$, $S \subseteq T$. Then $\xi_n = \varphi_n(\xi)$ also has this property. In fact,

$$\{\xi_n \subseteq S\} = \bigcup_{x \subseteq S} \{\xi_n \subseteq x\},$$

the union being taken over all $x \in X_n$, and

$$\{\xi_n \subseteq x\} = \{\xi \subseteq x\},$$

ξ_n is the minimal set in X_n containing ξ, so that $\xi_n \subseteq x$ if $x \supseteq \xi$. In short,

$$\{\xi_n \subseteq S\} = \bigcup_{x \subseteq S} \{\xi \subseteq x\} \in \mathscr{A}(S)^\varepsilon, \qquad \varepsilon > 0.$$

Clearly, the compatibility condition for $\xi_n \in X_n$ implies that

$$\{\xi_n \subseteq x\} \in \mathscr{A}(x)^\varepsilon, \qquad \varepsilon > 0, \qquad x \in X_n.$$

We will show that the compatibility of ξ_n is equivalent to the condition

$$\{\xi_n = x\} \in \mathscr{A}(x)^\varepsilon, \qquad \varepsilon > 0, x \in X_n. \tag{2.30}$$

It is clear that the compatibility of ξ_n is implied by the following condition:

$$\{\xi_n \subseteq x\} = \bigcup_{y \subseteq x} \{\xi_n = y\} \in \bigvee_{y \subseteq x} \mathscr{A}(y)^\varepsilon \subseteq \mathscr{A}(x)^\varepsilon$$

(the union is taken over $y \in X_n$). On the other hand, we have

$$\{\xi_n = x\} = \{\xi_n \subseteq x\} \setminus \bigcup_{y \subset x} \{\xi_n \subseteq x, \xi_n \not\subseteq y\} \in \mathscr{A}(x)^\varepsilon,$$

cf. (2.22).

We consider the σ-algebra $\mathscr{B}$ generated by events (2.29). For every open domain $S \subseteq T$ we have

$$\{\xi_n \subseteq S\} = \bigcup_{x \subseteq S} \{\xi \subseteq x\}, \qquad x \in X_n,$$

and

$$\{\xi \subseteq S\} = \bigcup_n \{\xi_n \subseteq S\},$$

because the compact $\xi \subseteq S$ is a positive distance from the boundary of the open domain S and ξ_n is contained in S for sufficiently large n. Thus we have established the following:

$\mathscr{B}$ *is the smallest σ-algebra for which the random domain ξ is measurable.*

The following proposition is also true.

Theorem 2. *For a compact random domain ξ which is compatible with the family $\mathscr{A}(S)$, $S \subseteq T$, the stopping σ-algebra $\mathscr{A}(\xi)^\varepsilon$ is the collection of all events A such that*

$$A \cdot \{\xi \subseteq S\} \in \mathscr{A}(S)^\varepsilon \tag{2.31}$$

for every compact domain $S \subseteq T$.

PROOF. We know that

$$\mathscr{A}_1(\xi) = \mathscr{A}(\xi)^\varepsilon = \bigcap_n \mathscr{A}(\xi_n)^\varepsilon,$$

where each σ-algebra $\mathscr{A}_1(\xi_n) = \mathscr{A}(\xi_n)^\varepsilon$ consists of all events A for which

$$A \cdot \{\xi_n = x\} \in \mathscr{A}(x)^\varepsilon, \qquad x \in X_n,$$

see (2.3). We will show that this is equivalent to

$$A \cdot \{\xi_n \subseteq x\} \in \mathscr{A}(x)^\varepsilon, \qquad x \in X_n.$$

Indeed, if the first condition is satisfied then

$$A \cdot \{\xi_n \subseteq x\} = \bigcup_{y \subseteq x} A \cdot \{\xi_n = y\} \in \mathscr{A}(x)^\varepsilon.$$

Conversely,

$$A \cdot \{\xi_n = x\} = A \cdot \{\xi_n \subseteq x\} \setminus \bigcup_{y \subset x} A \cdot \{\xi_n \subseteq y\} \in \mathscr{A}(x)^\varepsilon,$$

if the second condition holds. From this, together with the equality

$$\{\xi \subseteq x\} = \{\xi_n \subseteq x\}, \qquad x \in X_n, n = 1, 2, \dots,$$

we get that $A \in \bigcap_n \mathscr{A}(\xi_n)^\varepsilon$ if and only if inclusion (2.31) is true for all $S \in X_n$, $n = 1, 2, \dots$. But for every compact domain $S \subseteq T$ we have $S = \bigcap_n S_n$ and

$$A \cdot \{\xi \subseteq S\} = \bigcap_n A \cdot \{\xi \subseteq S_n\} \in \mathscr{A}(S)^\varepsilon, \qquad \varepsilon > 0.$$

The proof is complete.   □

From this we see that the stopping σ-algebra $\mathscr{A}(\xi)^\varepsilon$ does not depend on the partitions (2.25). One might think that the other σ-algebras in (2.28) depend substantially on the choice of the sequence of partitions but this is not so.

We can relinquish the condition of compatibility with the family $\mathscr{A}(S)$, $S \subseteq T$, which has allowed us to give a description of the stopping σ-algebra $\mathscr{A}_1(\xi) = \mathscr{A}(\xi)^\varepsilon$ independent of the partitions (2.25). We look at a closed random domain η measurable with respect to the σ-algebra of events $\mathscr{B}$ and such that $\eta \subseteq \xi$ with probability 1. We will compare the stopping σ-algebra $\mathscr{A}(\eta)^\delta$, constructed from a collection of arbitrary partitions of the form (2.25) with cells y_{jm} of diameter $\leq \delta_m \to 0$, with the σ-algebra $\mathscr{A}(\xi)^\varepsilon$. We will show that

$$\mathscr{A}(\eta)^\delta \subseteq \mathscr{A}(\xi)^\varepsilon, \qquad \delta < \varepsilon. \tag{2.32}$$

As we know, the σ-algebra $\mathscr{A}(\xi_n)^\varepsilon$ consists of all those events A such that

$$A \cdot \{\xi_n = x\} = A(x^\varepsilon) \cdot \{\xi_n = x\},$$

where $A(x^\varepsilon)$ is an event in the σ-algebra $\mathscr{A}(x)^\varepsilon = \mathscr{A}(x^\varepsilon)$. The σ-algebra $\mathscr{A}(\eta_m)^\delta$ has a similar structure. For an arbitrary $A \in \mathscr{A}(\eta_m)^\delta$ we have

$$A \cdot \{\xi_n = x\} = \bigcup_{y \subseteq x^{\delta m}} [A \cdot \{\eta_m = y\}] \cdot \{\xi_n = x\}$$

because

$$\{\xi_n = x\} = \bigcup_y \{\xi_n = x\} \cdot \{\eta_m = y\},$$

where the intersections on the right are nonempty only in the event that $y \subseteq x^{\delta m}$; but η_m consists of cells of diameter $\leq \delta_m$ which impinge on the domain η, and for $\eta \subseteq \xi$ we have $\eta_m \subseteq \eta^{\delta m} \subseteq \xi^{\delta m} \subseteq \xi_n^{\delta m}$. As we observed earlier, for $A \in \mathscr{A}(\eta_m)^\delta$ we have

$$A \cdot \{\eta_m = y\} = A(y^\delta) \{\eta_m = y\},$$

where $A(y^\delta)$ is an event in the σ-algebra $\mathscr{A}(y)^\delta = \mathscr{A}(y^\delta)$, and thus for $y^\delta \subseteq x^\varepsilon$ (for instance, for $\delta_m + \delta < \varepsilon$)

$$A(y^\delta) \in \mathscr{A}(x)^\varepsilon, \qquad A(y^\delta) \cdot \{\xi_n = x\} \in \mathscr{A}(\xi_n)^\varepsilon.$$

Assuming that all the random domains η_m are measurable with respect to the σ-algebra $\mathscr{B}$ we get

$$A \cdot \{\xi_n = x\} = \bigcup_y [A(y^\delta)\{\xi_n = x\}] \cdot \{\eta_m = y\} \in \mathscr{A}(\xi_n)^\varepsilon \vee \mathscr{B}$$

and

$$A = \bigcup_x [A \cdot \{\xi_n = x\}] \in \mathscr{A}(\xi_n^\varepsilon) \vee \mathscr{B}.$$

Thus

$$\mathscr{A}(\eta_m)^\delta \vee \mathscr{B} \subseteq \mathscr{A}(\xi_n)^\varepsilon \vee \mathscr{B}, \qquad \delta_m + \delta < \varepsilon,$$

where $\delta_m \to 0$ as $m \to \infty$; as a result,

$$\mathscr{A}(\eta)^\delta = \bigcap_m [\mathscr{A}(\eta_m)^\delta \vee \mathscr{B}] \subseteq \mathscr{A}(\xi_n)^\varepsilon \vee \mathscr{B}, \qquad \delta < \varepsilon,$$

and (2.32) follows.

In the same fashion we get similar relations involving $\mathscr{B}(\xi) = \mathscr{A}(\partial\xi)^\varepsilon$ and $\mathscr{A}_2(\xi) = \mathscr{A}(\xi^c)^\varepsilon$. We now state a summarizing result for all these sampling σ-algebras.

Theorem 3. *Let $\mathscr{B}$ be the minimal σ-algebra of events with respect to which the random domain ξ is measurable and let the random domain η be measurable with respect to $\mathscr{B}$. Then for any $\delta < \varepsilon$*

$$
\begin{aligned}
\mathscr{A}(\eta)^\delta &\subseteq \mathscr{A}(\xi)^\varepsilon, & \eta &\subseteq \xi, \\
\mathscr{A}(\partial\eta)^\delta &\subseteq \mathscr{A}(\partial\xi)^\varepsilon, & \eta &= \xi, \\
\mathscr{A}(\eta^c)^\delta &\subseteq \mathscr{A}(\xi^c)^\varepsilon, & \eta &\supseteq \xi,
\end{aligned}
\qquad (2.33)
$$

where the relationships $\eta \subseteq \xi, \eta = \xi, \eta \supseteq \xi$ hold with probability 1.

The inclusions (2.33) show that roughly speaking, for $\eta = \xi$ our stopping σ-algebras (2.28) do not depend on the sequence of partitions used to determine them. This statement is precisely correct for

$$
\begin{aligned}
\mathscr{A}_+(\xi) &= \bigcap_{\varepsilon > 0} \mathscr{A}(\xi)^\varepsilon, \\
\mathscr{A}_+(\partial\xi) &= \bigcap_{\varepsilon > 0} \mathscr{A}(\partial\xi)^\varepsilon, \\
\mathscr{A}_+(\xi^c) &= \bigcap_{\varepsilon > 0} \mathscr{A}(\xi^c)^\varepsilon.
\end{aligned}
\qquad (2.34)
$$

As we have in fact seen, by lemma 2, the stopping σ-algebras (2.28) are a Markov sequence for any sufficiently small $\varepsilon > 0$—i.e., the σ-algebra

$\mathscr{A}(\partial\xi)^{\varepsilon}$ splits $\mathscr{A}(\xi)^{\varepsilon}$ and $\mathscr{A}(\xi^c)^{\varepsilon}$. We call this the *strong Markov property for the family* $\mathscr{A}(S)$, $S \subseteq T$.

We state our main result in the following theorem.

Theorem 4. *The random field* $\mathscr{A}(S)$, $S \subseteq T$, *which is Markov for all relatively compact domains* $S \subseteq T$, *has the strong Markov property with respect to all compact random domains* $\xi \subseteq T$ *compatible with the field.*

§3. Gaussian Fields. Markov Behavior in the Wide Sense

1. Gaussian Random Fields

Let a random field $\mathscr{A}(S)$, $S \subseteq T$, be such that each σ-algebra $\mathscr{A}(S)$ is generated by variables in some corresponding space

$$H(S) \subseteq L^2(\Omega, \mathscr{A}, P)$$

and the family $H(S)$, $S \subseteq T$, defined on the system of open dmains $S \subseteq T$, is additive:

$$H(S' \cup S'') = H(S') \vee H(S'') \tag{3.1}$$

for every $S', S'' \subseteq T$. We call this collection of spaces $H(S)$, $S \subseteq T$, having the additivity property (3.1) a *random field* also.

We call the random field $H(S)$, $S \subseteq T$, *continuous* if

$$H(S) = \bigvee_n H(S^{(n)}), \qquad S = \bigcup S^{(n)}, \tag{3.2}$$

for any sequence of monotonically increasing domains $S^{(n)}$, $n = 1, 2, \ldots$; here and in what follows, the symbol $\bigvee$ denotes the closed linear span of the spaces indicated.

We call the random field *Gaussian* if the spaces $H(S)$, $S \subseteq T$, are formed by Gaussian random variables with zero mean.

EXAMPLE. Let $\xi(t)$, $t \in T$, be a Gaussian random function on a space T and

$$H(S) = \bigvee_{t \in S} \xi(t)$$

be the closed linear space of the variables $\xi(t)$, $t \in S$. The family $H(S)$, $S \subseteq T$, is a continuous Gaussian field.

EXAMPLE. Let (u, ξ), $u \in C_0(T)$, be a generalized Gaussian function on a domain $T \subseteq \mathbb{R}^d$ and

$$H(S) = \bigvee_{\mathrm{Supp}\, u \subseteq S} (u, \xi)$$

be the closed linear span of the variables (u, ξ) with Supp $u \subseteq S$, a domain. The family $H(S)$, $S \subseteq T$, is a continuous Gaussian field—see the example on p. 63.

We consider an arbitrary Gaussian field $H(S)$, $S \subseteq T$. By our definition, it is Markov with respect to some system of domains $S \in \mathcal{G}$ if for domains $S_1 = S$ and $S_2 = T \backslash \bar{S}$, and $\Gamma \supseteq \partial S$ the boundary separating them, the spaces $H(S_1)$ and $H(S_2)$ are conditionally independent with respect to $H(\Gamma^\varepsilon)$ for every sufficiently small neighborhood Γ^ε of the boundary Γ, cf. (1.26).

Let $P(H)$ be the orthogonal projection operator on the subspace $H \subseteq H(T)$, letting

$$P(S) = P(H(S))$$

for short.

The conditional distributions of the Gaussian variables $\eta \in H(T)$ with respect to $H \subseteq H(T)$ are Gaussian with conditional expectations

$$E(\eta \mid H) = P(H)\eta$$

and covariances

$$E[\eta_1 - P(H)\eta_1][\eta_2 - P(H)\eta_2], \qquad \eta_1, \eta_2 \in H(T).$$

In this context, the conditional independence of the subspaces H_1 and H_2 with respect to H is equivalent to every variable $\eta_1 \in H_1$ and $\eta_2 \in H_2$ being conditionally uncorrelated, that is,

$$\eta_1 - P(H)\eta_1 \perp \eta_2 - P(H)\eta_2, \qquad \eta_i \in H_i. \tag{3.3}$$

Thus the Markov property for a Gaussian field $H(S)$, $S \subseteq T$, is equivalent to the orthogonality condition (3.3) holding for any domain $S \in \mathcal{G}$ and corresponding

$$H_1 = H(S_1), \qquad H = H(\Gamma^\varepsilon), \qquad H_2 = H(S_2). \tag{3.4}$$

2. Splitting Spaces

Let H be a subspace of the Hilbert space $L^2(\Omega, \mathcal{A}, P)$ and $P(H)$ be the orthogonal projection operator on H. We denote by $H_1 \ominus H$ the subspace generated by variables $\eta - P(H)\eta$, $\eta \in H_1$. We call the space H *splitting* for H_1 and H_2 if $H_1 \ominus H$ and $H_2 \ominus H$ are orthogonal: condition (3.3) holds for $\eta_1 \in H_1$ and $\eta_2 \in H_2$.

The space H being splitting is equivalent to each of the relations

$$(H_1 \ominus H) \perp (H_2 \vee H), \qquad (H_2 \ominus H) \perp (H_1 \vee H), \tag{3.5}$$

because

$$(H_2 \ominus H) \vee H \supseteq H_2, \qquad (H_1 \ominus H) \vee H \supseteq H_1;$$

moreover, the space H is splitting for H_1, H_2 and simultaneously for $H_1 \vee H$, $H_2 \vee H$.

H being splitting means that

$$H_1 \vee H \vee H_2 = (H_1 \ominus H) \oplus H \oplus (H_2 \ominus H), \qquad (3.6)$$

where the symbol $\oplus$ denotes the orthogonal sum of the spaces and

$$(H_1 \ominus H) \oplus H = H_1 \vee H, \qquad (H_2 \ominus H) \oplus H = H_2 \vee H.$$

In the case where $H \subseteq H_1$, H_2, equation (3.6) becomes

$$H_1 \cap H_2 = H, \qquad H_1^\perp \perp H_2^\perp, \qquad (3.7)$$

where $H_i^\perp$ is the orthogonal complement of H_i in the space $H_1 \vee H_2$.

Here are some other properties of splitting spaces.

Suppose the spaces H', H'' are contained in H_1 and are splitting for H_1, H_2; then the intersection $H = H' \cap H''$ is splitting for H_1, H_2.

This follows because for $H \subseteq H_1$, (3.6) is equivalent to the inclusion

$$P(H_1)\eta \in H \quad \text{for } \eta \in H_2$$

and this is clearly satisfied.

If $H^{(n)}$, $n = 1, 2, \ldots$, is a monotonically decreasing sequence of spaces which split H_1 and H_2, then their intersection $H = \bigcap H^{(n)}$ also splits H_1, H_2.

In fact, for such a sequence

$$\eta - P(H)\eta = \lim_{n \to \infty} [\eta - P(H^{(n)})\eta],$$

and for $\eta \in H_1$, all the variables $\eta - P(H^{(n)})\eta$ are orthogonal to $H_2 \vee H^{(n)}$, and the intersection $H = \bigcap H^{(n)}$ satisfies condition (3.5).

If $H^{(n)} \subseteq H_1$ is any sequence of spaces which split H_1, H_2, then the intersection $H = \bigcap H^{(n)}$ splits H_1, H_2. Among all splitting spaces $H \subseteq H_1$ there is a minimal one, H_0, and by (3.6) we have

$$H_0 \supseteq (H_1 \cap H_2), \qquad (3.8)$$

since by (3.5) the intersection $H_1 \cap H_2$ is orthogonal to $H_1 \ominus H_0$ and to $H_2 \ominus H_0$.

Let the space H_0 split H_1, H_2. Generally speaking, a larger space $H \supseteq H_0$ will not be splitting. However, if

$$H = H' \vee H_0 \vee H'', \qquad (3.9)$$

with $H' \subseteq H_1$ and $H'' \subseteq H_2$, then H is also splitting.

To see this, observe that H_0 will be splitting for $H_1 \vee H_0$ and $H_2 \vee H_0$, so that for $\eta \in H_1 \vee H_0$ the difference

$$\eta - P(H' \vee H_0)\eta = [\eta - P(H_0)\eta] - [P(H' \vee H_0)\eta - P(H_0)P(H' \vee H_0)\eta],$$

where

$$P(H' \vee H_0)\eta \in H_1 \vee H_0$$

is orthogonal to $H' \vee H_0$ and $H_2 \vee H_0$—that is, $H' \vee H_0$ will split $H_1 \vee H_0$ and $H_2 \vee H_0$, in view of (3.5). By the same considerations, the space $H = (H' \vee H_0) \vee H''$ also will be splitting.

3. Markov Property

We consider a random field $H(S)$, $S \subseteq T$, on a locally-compact separable metric space T, and a system of open domains $S \in \mathscr{G}$, with each set S having an associated triple

$$S_1 = S, \qquad \Gamma \supseteq \partial S, \qquad S_2 = T\backslash(S \cup \Gamma),$$

see (1.26). We call the random field *Markov* (*in the wide sense*) with respect to the system $\mathscr{G}$ if for every domain $S \in \mathscr{G}$ and for every sufficiently small neighborhood Γ^ε of the boundary Γ, the subspace $H(\Gamma^\varepsilon)$ splits $H(S_1)$ and $H(S_2)$. In what follows we will consider this particular form of the Markov property, which for Gaussian fields implies the Markov property for the random field $\mathscr{A}(S)$, $S \subseteq T$, with the σ-algebras $\mathscr{A}(S)$ generated by random variables in the corresponding spaces $H(S)$, $S \subseteq T$.

We will say that a *domain S splits* domains S_1 and S_2 if $H(S)$ splits $H(S_1)$ and $H(S_2)$. We remark that if a domain S_0 splits S_1 and S_2, then each domain $S \supseteq S_0$ satisfying the condition

$$S \subseteq (S_1 \cup S_0 \cup S_2),$$

also splits S_1 and S_2 because (see (3.1) and (3.9))

$$H(S) = H(S_1 \cap S) \vee H(S_0) \vee H(S \cap S_2).$$

For convenience we will call complementary domains S_1, S_2 with separating boundary Γ *Markov* if a sufficiently small neighborhood Γ^ε splits S_1 and S_2. Let S_1, S_2 be complementary Markov domains with separating boundary Γ. Then by the additivity of our random field, the neighborhood Γ^ε splits not only S_1 and S_2 but also $S_1 \cup \Gamma^\varepsilon$ and $S_2 \cup \Gamma^\varepsilon$; equivalently,

$$H(S_1 \cup \Gamma^\varepsilon) \cap H(S_2 \cup \Gamma^\varepsilon) = H(\Gamma^\varepsilon),$$

$$H(S_1 \cup \Gamma^\varepsilon)^\perp \perp H(S_2 \cup \Gamma^\varepsilon)^\perp,$$

$$(3.10)$$

because

$$H(S_1 \cup \Gamma^\varepsilon) \vee H(S_2 \cup \Gamma^\varepsilon) = H(T)$$

and $H^\perp$ denotes the orthogonal complement of H in $H(T)$.

The Markov property for domains S_1 and S_2 can be expressed by the equation

$$P(S_1 \cup \Gamma^\varepsilon)H(S_2 \cup \Gamma^\varepsilon) = H(\Gamma^\varepsilon). \tag{3.11}$$

This is because from (3.10) we get that

$$H(S_1 \cup \Gamma^\varepsilon) = H(\Gamma^\varepsilon) \oplus H(S_2 \cup \Gamma^\varepsilon)^\perp,$$

$$H(S_2 \cup \Gamma^\varepsilon) = H(\Gamma^\varepsilon) \oplus H(S_1 \cup \Gamma^\varepsilon)^\perp,$$

which gives (3.11); conversely (3.11) implies that the subspace $H(S_2 \cup \Gamma^\varepsilon)$ $\ominus H(\Gamma^\varepsilon)$ is orthogonal to $H(S_1 \cup \Gamma^\varepsilon)$ and $H(\Gamma^\varepsilon)$, i.e., $H(\Gamma^\varepsilon)$ splits $H(S_1 \cup \Gamma^\varepsilon)$, $H(S_2 \cup \Gamma^\varepsilon)$. It is obvious also that (3.11) is equivalent to its companion equation

$$P(S_2 \cup \Gamma^\varepsilon)H(S_1 \cup \Gamma^\varepsilon) = H(\Gamma^\varepsilon). \tag{3.11a}$$

Now we set

$$H_+(S) = \bigcap_{\varepsilon > 0} H(S^\varepsilon)$$

for $S \subseteq T$ and ε-neighborhoods S^ε of S and we consider spaces of the form (3.4).

In the case of Markov domains S_1, S_2, each space $H(\Gamma^\varepsilon)$ is splitting for $H_+(S_1 \cup \Gamma^\delta)$ and $H_+(S_2 \cup \Gamma^\delta)$ if $\varepsilon > \delta$. Thus the space

$$H_+(\Gamma^\delta) = \bigcap_{\varepsilon > \delta} H(\Gamma^\varepsilon)$$

will be splitting for $H_+(S_1 \cup \Gamma^\delta)$, $H_+(S_2 \cup \Gamma^\delta)$ and this is equivalent to conditions analogous to (3.10):

$$\begin{aligned} H_+(S_1 \cup \Gamma^\delta) \cap H_+(S_2 \cup \Gamma^\delta) &= H_+(\Gamma^\delta), \\ H_+(S_1 \cup \Gamma^\delta)^\perp &\perp H_+(S_2 \cup \Gamma^\delta)^\perp, \qquad \delta > 0, \end{aligned} \tag{3.12}$$

because for $\delta > 0$ the additivity of the field gives

$$H_+(S_1 \cup \Gamma^\delta) \vee H_+(S_2 \cup \Gamma^\delta) = H(T).$$

Conversely, if $H_+(\Gamma^\delta)$ splits $H_+(S_1 \cup \Gamma^\delta)$, $H_+(S_2 \cup \Gamma^\delta)$ for any small neighborhood Γ^δ, then for $\varepsilon > \delta$ the space

$$H(\Gamma^\varepsilon) = H[\Gamma^\varepsilon \cap (S_1 \cup \Gamma^\delta)] \vee H[\Gamma^\varepsilon \cap (S_2 \cup \Gamma^\delta)]$$

will split $H_+(S_1 \cup \Gamma^\delta)$ and $H_+(S_2 \cup \Gamma^\delta)$ and thus also

$$H(S_1 \cup \Gamma^\varepsilon) = H_+(S_1 \cup \Gamma^\delta) \vee H(\Gamma^\varepsilon),$$

$$H(S_2 \cup \Gamma^\varepsilon) = H_+(S_2 \cup \Gamma^\delta) \vee H(\Gamma^\varepsilon).$$

Hence the Markov conditions (3.10) and (3.12) are equivalent.

For Markov domains S_1, S_2, the intersection

$$H_+(\Gamma) = \bigcap_{\varepsilon > 0} H(\Gamma^\varepsilon)$$

is the minimal space which splits $H_+(S_1 \cup \Gamma)$ and $H_+(S_2 \cup \Gamma)$ since $H_+(\Gamma)$ is contained in $H_+(S_1 \cup \Gamma)$ and in $H_+(S_2 \cup \Gamma)$ (cf. (3.8)). We will assume that

$$H(T) = H_+(S_1 \cup \Gamma) \vee H_+(S_2 \cup \Gamma); \tag{3.13}$$

§3. Gaussian Fields. Markov Behavior in the Wide Sense

97

then by (3.7), the space $H_+(\Gamma)$ splits $H_+(S_1 \cup \Gamma)$, $H_+(S_2 \cup \Gamma)$ if and only if

$$H_+(\Gamma) = H_+(S_1 \cup \Gamma) \cap H_+(S_2 \cup \Gamma),$$
$$H_+(S_1 \cup \Gamma)^\perp \perp H_+(S_2 \cup \Gamma)^\perp. \tag{3.14}$$

In general, conditions (3.14) are weaker than (3.10), (3.12) but they are equivalent under the assumption that for a neighborhood Γ^ε the following holds:

$$H(\Gamma^\varepsilon) = H(\Gamma^\varepsilon \cap S_1) \vee H_+(\Gamma) \vee H(\Gamma^\varepsilon \cap S_2). \tag{3.15}$$

In fact, by (3.9), if $H_+(\Gamma)$ splits $H_+(S_1 \cup \Gamma) \supseteq H(S_1)$ and $H_+(S_2 \cup \Gamma)$ $\supseteq H(S_2)$, then the space $H(\Gamma^\varepsilon)$ will split $H(S_1)$ and $H(S_2)$. As we did for (3.10) and (3.11), we can express (3.14) by the equation

$$P(H_+(S_1 \cup \Gamma))H_+(S_2 \cup \Gamma) = H_+(\Gamma) \tag{3.16}$$

or the symmetric version

$$P(H_+(S_2 \cup \Gamma))H_+(S_1 \cup \Gamma) = H_+(\Gamma). \tag{3.16a}$$

We will call a random field $H(S)$, $S \subseteq T$, *Markov* if it is Markov in a wide sense with respect to a complete system $\mathcal{G}$ of open domains $S \subseteq T$ and the middle term of the triple associated to S is $\Gamma = \partial S$, the topological boundary between S and $S_1 = S$ and $S_2 = T \backslash \bar{S}$.

As we did in deriving the Markov behavior of the sequence (1.31), we can confine ourselves to the Markov field situation where for each domain S, the one-sided neighborhood

$$\Gamma^\varepsilon_- = S \cap \Gamma^\varepsilon$$

splits $S_1 = S$ and $S_2 = T \backslash \bar{S}$ and the intersection

$$H_+(\Gamma_-) = \bigcap_{\varepsilon > 0} H(\Gamma^\varepsilon_-)$$

splits $H(S_1)$ and $H_+(S_2 \cup \Gamma) \supseteq H(S_2)$. Moreover, if the following analog of (1.32)

$$H(\Gamma^\varepsilon) = H(S_1 \cap \Gamma^\varepsilon) \vee H(S_2 \cap \Gamma^\varepsilon) \tag{3.17}$$

holds, then whether or not the random field is Markov is equivalent to whether or not $H_+(\Gamma_-)$ splits $H(S_1)$ and $H(S_2)$; this is because under (3.17), $H(\Gamma^\varepsilon) \supseteq H_+(\Gamma_-)$ splits $H(S_1)$ and $H(S_2)$ whenever $H_+(\Gamma_-)$ does. In particular, (3.17) is true for the random field

$$H(S) = \bigvee_{t \in S} \xi(t), \qquad S \subseteq T, \tag{3.18}$$

generated by continuous functions $\xi(t)$, $t \in T$, in $L^2(\Omega, \mathcal{A}, P)$.

Note that if a random field is exterior continuous, i.e.,

$$H_+(S) = \bigcap_{\varepsilon > 0} H(S^\varepsilon) = H(S), \tag{3.19}$$

we have the following analog of (1.34):

$$H_+(\Gamma_-) = H_+(\Gamma). \tag{3.20}$$

We can show that for a Markov field satisfying (3.17) conditions (3.19) and (3.20) are the same. From the representation

$$H(S_1 \cup \Gamma^\varepsilon) = H(S_1) \vee H(S_2 \cap \Gamma^\varepsilon) = H(S_1) \oplus [H(S_2 \cap \Gamma^\varepsilon) \ominus H_+(\Gamma_-)],$$
$$S_1 = S.$$

and equation (3.20), it follows that

$$H_+(S_1) = \bigcap_{\varepsilon > 0} H(S_1 \cup \Gamma^\varepsilon) = H(S_1),$$

since from the definition of $H_+(\Gamma)$ it is always true that

$$\bigcap_{\varepsilon > 0} [H(S_2 \cap \Gamma^\varepsilon) \ominus H_+(\Gamma)] \subseteq \bigcap_{\varepsilon > 0} [H(\Gamma^\varepsilon) \ominus H_+(\Gamma)] = 0.$$

4. Orthogonal Random Fields

We call a random field $H(S)$, $S \subseteq T$, *orthogonal* with respect to a system $\mathcal{G}_0$ of open domains $S \subseteq T$ if for $S_1 = S$, $S_2 = T \backslash \bar{S}$ and a boundary $\Gamma \supseteq \partial S$ we have

$$H(S_1 \backslash \Gamma) \perp H_2(S_2 \backslash \Gamma). \tag{3.21}$$

Clearly, if points t_1, $t_2 \in T$ are separated by the boundary Γ so that $t_1 \in S_1 \backslash \Gamma$ and $t_2 \in S_2 \backslash \Gamma$, then for an orthogonal random field one can find neighborhoods t_1^ε and t_2^ε such that

$$H(t_1^\varepsilon) \perp H(t_2^\varepsilon), \qquad t_1 \neq t_2; \tag{3.22}$$

moreover, condition (3.22) holding for sufficiently small neighborhoods implies (3.21).

In fact, any open $S \subseteq T$ can be represented as the union $S = \bigcup_n S^{(n)}$ of some sequence of compact domains $S^{(n)}$. Each of the compact sets $(S_1 \backslash \Gamma)^{(n)}$, $(S_2 / \Gamma)^{(n)}$ has a finite covering of neighborhoods $t_1^\varepsilon \subset S_1$ and $t_2^\varepsilon \subset S_2$ and the spaces $H(t_1^\varepsilon)$ and $H(t_2^\varepsilon)$ are orthogonal. By the additivity and continuity of the random field, the spaces $H(S_1 \backslash \Gamma) = \vee H(t_1^\varepsilon)$ and $H(S_2 \backslash \Gamma) = \vee H(t_2^\varepsilon)$ are also orthogonal.

We have thus shown the following.

Let a system of domains $S \in \mathcal{G}_0$ have the property that the domains $S_1 \backslash \Gamma$, $S_2 \backslash \Gamma$ separate points. Then a continuous random field which is orthogonal with respect to the system $\mathcal{G}_0$ is also orthogonal with respect to all open domains $S \subseteq T$ with boundaries $\Gamma = \partial S$; that is,

$$H(S_1) \perp H(S_2), \qquad S_1 = S, S_2 = T \backslash \bar{S}, \tag{3.23}$$

for any open domain $S \subseteq T$.

5. Dual Fields. A Markov Criterion

Let $\mathscr{G}$ be a complete system of open domains $S \subseteq T$. We say a random field $H^*(S)$, $S \subseteq T$, is *dual* (or *conjugate*) to the random field $H(S)$, $S \subseteq T$, *on the system $\mathscr{G}$* if

$$H^*(T) = H(T)$$

and

$$H^*(S) = H_+(S^c)^\perp \quad \text{for every domain } S \in \mathscr{G}, \tag{3.24}$$

where $S^c = T \backslash S$ and $\perp$ denotes the usual orthogonal complement in the space $H(T)$.

Let S', $S'' \in \mathscr{G}$; under the duality condition (3.24) we have the following equation for the complements $F' = T \backslash S'$ and $F'' = T \backslash S''$:

$$H_+(F') \cap H_+(F'') = H_+(F' \cap F''). \tag{3.25}$$

In fact,

$$H_+(F') \cap H_+(F'') = H^*(S')^\perp \cap H^*(S'')^\perp = [H^*(S') \vee H^*(S'')]^\perp$$
$$= H^*(S' \cup S'')^\perp = H_+(F' \cap F'').$$

In the discussion below we shall make use of the fact that for any sub-spaces H_k, $k = 1, 2, \ldots$,

$$(\cap H_k)^\perp = \bigvee_k H_k^\perp. \tag{3.26}$$

We note that a continuous random field $H(S)$, $S \subseteq T$, is itself dual to its dual field:

$$H(S) = H_+^*(S^c)^\perp. \tag{3.27}$$

This is true because for the domains

$$S^{-\varepsilon} = T \backslash \overline{(S^c)^\varepsilon}, \quad \varepsilon > 0,$$

we have

$$H(S) = \bigvee_{\varepsilon > 0} H_+(S^{-\varepsilon}),$$

where

$$H_+(S^{-\varepsilon}) = H^*[(S^c)^\varepsilon]^\perp,$$

$$\bigvee_{\varepsilon > 0} H^*[(S^c)^\varepsilon]^\perp = \left\{ \bigcap_{\varepsilon > 0} H^*[(S^c)^\varepsilon] \right\}^\perp = H_+^*(S^c)^\perp.$$

Let $H^*(S)$, $S \subseteq T$, be a dual random field. For

$$F' = \overline{S_1 \cup \Gamma^\varepsilon} = S_1 \cup \overline{\Gamma^\varepsilon}, \qquad F'' = \overline{S_2 \cup \Gamma^\varepsilon} = S_2 \cup \overline{\Gamma^\varepsilon},$$

equation (3.25) gives the first of the Markov conditions (3.12) for the random field $H(S)$, $S \subseteq T$, namely,

$$H_+(S_1 \cup \Gamma^\varepsilon) \cap H_+(S_2 \cup \Gamma^\varepsilon) = H_+(\Gamma^\varepsilon)$$

for complementary domains $S_1 = S$ and $S_2 = T \backslash \bar{S}$ and $\Gamma \supset \partial S$ the boundary separating them. The second of these conditions can be expressed in the following fashion:

$$H^*(S_1 \backslash \Gamma) \perp H^*(S_2 \backslash \Gamma). \tag{3.28}$$

This follows from

$$(S_1 \backslash \Gamma)^c = S_2 \cup \Gamma, \qquad (S_2 \backslash \Gamma)^c = S_1 \cup \Gamma$$

and

$$H_+(S_1 \cup \Gamma)^\perp = H^*(S_1 \backslash \Gamma), \qquad H_+(S_2 \cup \Gamma)^\perp = H^*(S_2 \backslash \Gamma);$$

see the second of the conditions (3.14). The orthogonality relation in (3.28) exactly implies the orthogonality of the dual random field with respect to the collection of domains $S \in \mathcal{G}_0$. Here $\mathcal{G}_0 \subseteq \mathcal{G}$ is the system with respect to which we are considering the Markov property of the field $H(S)$, $S \subseteq T$.

With the results of §3.4 in mind, the existence of the dual field on the complete system $\mathcal{G}$ gives us a criterion for the Markov property; namely, we have the following.

Theorem. *A random field $H(S)$, $S \subseteq T$, is Markov with respect to some system $\mathcal{G}_0 \subseteq \mathcal{G}$ if and only if the dual field is orthogonal with respect to $\mathcal{G}_0$. In the case where $\mathcal{G}_0$ separates points $t \in T$, the random field $H(S)$, $S \subseteq T$, is Markov with respect to $\mathcal{G}$.*

6. Regularity Condition. Decomposition of a Markov Field into Regular and Singular Components

Let $H(S)$, $S \subseteq T$, be a random field. We will study its behavior on domains $S^{(n)}$, $n = 1, 2, \ldots$, forming a complete neighborhood system of a "point at infinity"; more precisely, the $S^{(n)}$ are complements of a collection of compact sets $F^{(n)}$, $n = 1, 2, \ldots$, such that every compact $F \subseteq T$ is contained in some $F^{(n)}$. (Recall that T is a locally-compact separable metric space.) We call the random field *regular*† if

$$\bigcap_n H(S^{(n)}) = 0 \tag{3.29}$$

---

† For a Gaussian random field, (3.29) is equivalent to the regularity condition for $\mathcal{A}(S)$, $S \subseteq T$, where the σ-algebras of events $\mathcal{A}(S)$ are generated by the corresponding spaces of Gaussian variables $H(S)$, $S \subseteq T$.

and *singular* if

$$\bigcap_n H(S^{(n)}) = H(T). \tag{3.30}$$

Notice that the regularity condition (3.29) is satisfied if there exists a continuous conjugate field $H^*(S)$, $S \subseteq T$, since

$$\bigcap_n H(S^{(n)}) = \bigcap_n H_+^*(F^{(n)})^\perp = \left[\bigvee_n H_+^*(F^{(n)}) \right]^\perp$$

$$= H^*(T)^\perp = H(T)^\perp = 0.$$

Let $P = P(\infty)$ be the orthogonal projection on the subspace

$$H(\infty) = \bigcap_n H(S^{(n)}).$$

The family

$$H_1(S) = \overline{(I - P)H(S)}, \qquad S \subseteq T, \tag{3.31}$$

forms a regular field, so that

$$H_2(S) = \overline{PH(S)}, \qquad S \subseteq T, \tag{3.32}$$

is a singular random field. (As usual, a line over a set denotes its closure.)

Moreover, together with the random field $H(S)$, $S \subseteq T$, these fields are additive and continuous. We have

$$H(S^{(n)}) = H(\infty) \oplus H_1(S^{(n)}),$$

$$H_2(S^{(n)}) = PH(S^{(n)}) = H(\infty), \qquad n = 1, 2, \ldots,$$

and

$$\bigcap_n H_1(S^{(n)}) = \bigcap_n [H(S^{(n)}) \ominus H(\infty)] = 0.$$

We call the random fields (3.31) and (3.32) the *regular* and *singular components* of the random field.

We will show that if the random field is Markov then its regular and singular components also have this property. We stress that this pertains to the Markov property with respect to the system of open domains $S \in \mathscr{G}$ which are relatively compact or have compact complements, with boundaries $\Gamma \supseteq \partial S$ between the domains $S_1 = S$ and $S_2 = T \backslash \bar{S}$.

By virtue of the symmetry of the Markov property with respect to S_1 and S_2, we can take for $S_1 = S$ a domain which is the complement of a compact set. Such a domain contains a neighborhood $S^{(n)}$ of the "point at infinity" for n sufficiently large, so that we have

$$H(S_1 \cup \Gamma^\varepsilon) = H_1(S_1 \cup \Gamma^\varepsilon) \oplus H_2(S_1 \cup \Gamma^\varepsilon), \qquad H_2(S_1 \cup \Gamma^\varepsilon) = H(\infty)$$

and

$$P_1(S_1 \cup \Gamma^\varepsilon) = P(S_1 \cup \Gamma^\varepsilon) \cdot (I - P), \qquad P_2(S_1 \cup \Gamma^\varepsilon) = P(S_1 \cup \Gamma^\varepsilon) \cdot P$$

are the projection operators on the subspaces $H_1(S_1 \cup \Gamma^\varepsilon)$, $H_2(S_1 \cup \Gamma^\varepsilon)$. For all $\varepsilon > 0$ the inclusion

$$P(S_1 \cup \Gamma^\varepsilon)H(S_2 \cup \Gamma^\varepsilon) \subseteq H(\Gamma^\varepsilon)$$

gives us

$$(I - P) \cdot P(S_1 \cup \Gamma^\varepsilon)H(S_2 \cup \Gamma^\varepsilon) = P(S_1 \cup \Gamma^\varepsilon) \cdot (I - P)H(S_2 \cup \Gamma^\varepsilon)$$
$$= P_1(S_1 \cup \Gamma^\varepsilon)H_1(S_2 \cup \Gamma^\varepsilon)$$
$$\subseteq \overline{(I - P)H(\Gamma^\varepsilon)} = H_1(\Gamma^\varepsilon)$$

and similarly

$$P_2(S_1 \cup \Gamma^\varepsilon)H_2(S_2 \cup \Gamma^\varepsilon) \subseteq H_2(\Gamma^\varepsilon),$$

which is equivalent to the Markov condition.

We can state the result above as a proposition:

The random field $H(S)$, $S \subseteq T$, can be decomposed into regular and singular component Markov fields.

CHAPTER 3

The Markov Property for Generalized Random Functions

§1. Biorthogonal Generalized Functions and the Duality Property

1. The Meaning of Biorthogonality for Generalized Functions in Hilbert Space

We have previously stipulated that by a generalized random function on a domain $T \subseteq \mathbb{R}^d$ we mean a continuous linear mapping of $C_0^\infty(T)$, the space of infinitely differentiable functions $u = u(t)$, $t \in T$, into the Hilbert space $L^2(\Omega, \mathscr{A}, P)$.

In studying generalized random functions it is convenient to introduce the continuous linear operator J inducing the mapping $C_0^\infty(T) \to L^2(\Omega, \mathscr{A}, P)$, namely

$$(u, \xi) = Ju, \qquad u \in C_0^\infty(T). \tag{1.1}$$

With such a generalized function we associate the random field

$$H(S), \qquad S \subseteq T, \tag{1.2}$$

where for every open domain $S \subseteq T$, $H(S)$ is the closed linear span of the variables

$$(u, \xi), \qquad u \in C_0^\infty(S),$$

and $C_0^\infty(S)$ is regarded as the subspace of all functions $u \in C_0^\infty(T)$ with support Supp $u \subseteq S$.

We call the generalized random function

$$(u, \xi^*) = Lu, \qquad u \in C_0^\infty(T) \tag{1.3}$$

biorthogonal to the function (1.1) if

$$E(u, \xi) \cdot \overline{(v, \xi^*)} = (Ju, Lv) = \int_T u(t)\overline{v(t)}\, dt, \qquad u, v \in C_0^\infty(T) \tag{1.4}$$

and

$$H^*(T) = H(T); \tag{1.5}$$

here

$$H^*(S), \qquad S \subseteq T, \tag{1.6}$$

is the random field generated by the generalized function (1.3); that is, for any open domain $S \subseteq T$, $H^*(S)$ is the closed linear span of the variables (u, ξ^*), $u \in C_0^\infty(S)$.

Obviously, the biorthogonal function is unique, and the original function (1.1) is biorthogonal to its biorthogonal function (1.3).

The existence of a biorthogonal function can be demonstrated by regarding the space $C_0^\infty(T)$ as the domain of the associated operator J on the usual Hilbert space $L^2(T)$ of square-integrable functions $u(t)$, $t \in T$, with inner product

$$(u, v) = \int_T u(t)\overline{v(t)}\, dt, \qquad u, v \in L^2(T).$$

In particular, for the linear operator

$$J: C_0^\infty(T) \subseteq L^2(T) \to H(T),$$

conditions (1.4), (1.5) imply the existence of an adjoint operator J^* having a right inverse operator L with domain $C_0^\infty(T) \subseteq L^2(T)$,

$$J^*(Lv) = v, \qquad v \in C_0^\infty(T), \tag{1.7}$$

continuous on $C_0^\infty(T)$ and having a range which is dense in $H(T)$.

Now let us consider the so-called *reproducing kernel space* $V(T)$ for the random function (1.1); it consists of (deterministic) generalized functions

$$v = (u, v), \qquad u \in C_0^\infty(T),$$

—continuous linear functionals on the space $C_0^\infty(T)$—which can be represented by means of corresponding variables $\eta \in H(T)$ by

$$(u, v) = E(u, \xi)\eta = (Ju, \eta), \qquad u \in C_0(T). \tag{1.8}$$

It is clear that for each variable $\eta \in H(T)$ the formula (1.8) defines a generalized function because the linear operator J is continuous on $C_0^\infty(T)$. The correspondence

$$v = v(\eta) \leftrightarrow \eta = \eta(v) \tag{1.9}$$

is one-to-one, since the range of the operator J is dense in the space $H(T)$. If we introduce into the space $V(T)$ an inner product, defining it by the equation

$$\langle v_1, v_2 \rangle = (\eta(v_1), \eta(v_2)), \qquad v_i \in V(T), \tag{1.10}$$

then the mapping (1.9) is unitary.

Let us assume that we have a biorthogonal function (1.3). Then for each element $v \in C_0^\infty(T)$ the generalized function

$$v = (u, v) = \int_T u(t)\overline{v(t)}\, dt, \qquad u \in C_0^\infty(T), \tag{1.11}$$

belongs to the space $V(T)$ because according to (1.4) there is a variable $\eta(v) = Lv$ corresponding to v:

$$v \leftrightarrow Lv, \qquad v \in C_0^\infty(T). \tag{1.12}$$

Identifying the functions $v \in C_0^\infty(T)$ with generalized functions of the form (1.11), we can speak of an embedding

$$C_0^\infty(T) \subseteq V(T), \tag{1.13}$$

which is continuous and dense in the space $V(T)$, i.e., the norm

$$\|v\|_L = \|Lv\|, \qquad v \in C_0^\infty(T), \tag{1.14}$$

on the space $V(T)$ is continuous on $C_0^\infty(T)$ and the closure of $C_0^\infty(T)$ under this norm coincides with the whole space $V(T)$. The embedding (1.13) is a consequence of the continuity of the linear operator L and condition (1.5), according to which the variables $\eta = Lv$, $v \in C_0^\infty(T)$, are dense in the space $H(T)$, hence the corresponding generalized functions $v = v(\eta)$ are dense in the space $V(T)$. We can characterize the same space $V(T)$ as the closure of $C_0^\infty(T)$ under the norm (1.14), with scalar product

$$\langle v_1, v_2 \rangle = (Lv_1, Lv_2), \qquad v_1, v_2 \in C_0^\infty(T). \tag{1.15}$$

The embedding

$$C_0^\infty(T) \subseteq V(T), \qquad \overline{C_0^\infty(T)} = V(T)$$

occurs if and only if the biorthogonal function exists.

Indeed, from the indicated embedding one can define a generalized random function (1.3) by letting

$$Lv = \eta(v), \qquad v \in C_0^\infty(T),$$

where $\eta(v) \leftrightarrow v$ (cf. (1.8)). It will be biorthogonal for the function (1.1) because by the unitary correspondence (1.9) we have

$$(u, v) = (Ju, Lv), \qquad u \in C_0^\infty(T).$$

2. Duality of Biorthogonal Functions

We consider the random fields (1.2) and (1.6) generated by the biorthogonal generalized functions (1.1) and (1.3).

From (1.4) it follows that

$$H^*(S) \subseteq H_+(S^c)^\perp; \tag{1.16}$$

recall that $S^c = T \backslash S$ and for every set $S \subseteq T$ the associated space $H_+(S)$ is defined by

$$H_+(S) = \bigcap_{\varepsilon > 0} H(S^\varepsilon),$$

with the intersection taken over all ε-neighborhoods of the set S in T.

To show the inclusion (1.16) we take an arbitrary variable $\eta = Lu$, $u \in C_0^\infty(S)$. The support of u, being compact, is a positive distance from the set S^c, hence outside some neighborhood $(S^c)^\varepsilon$, so that

$$\int u(t)\overline{v(t)}\, dt = (Lu, Jv) = 0$$

for all functions $v \in C_0^\infty[(S^c)^\varepsilon]$ with support in the domain $(S^c)^\varepsilon$; this means that the variable $\eta = Lu$ is orthogonal to the space $H[(S^c)^\varepsilon]$ and thus to the space $H_+(S^c)$.

For any open domain $S \subseteq T$, we can describe the orthogonal complement $H(S)^\perp$ by turning to the reproducing kernel space $V(T)$ and by using, for $\eta \in H(T)$, the corresponding function $v = (u, v)$, $u \in C_0^\infty(T)$, from the space $V(T)$ (cf. (1.9)). In particular, the condition

$$\eta \perp H(S)$$

implies that

$$(Ju, \eta) = (u, v) = 0, \qquad u \in C_0^\infty(S);$$

that is, the support Supp v of the generalized function $v \in V(T)$ lies in the complement of the domain S.

Returning to the domain $(S^c)^\varepsilon$ we can say, keeping the isomorphism (1.9) in mind, that the orthogonal complement

$$H[(S^c)^\varepsilon]^\perp = \{\eta(v) \colon \text{Supp } v \subseteq \overline{S^{-\varepsilon}}\}$$

is unitarily isomorphic to the subspace of generalized functions $v \in V(T)$ with support in the closed domain $\overline{S^{-\varepsilon}} = T \backslash (S^c)^\varepsilon$. In the same way the orthogonal complement

$$H_+(S^c)^\perp = \bigvee \{\eta(v) \colon \text{Supp } v \subseteq \overline{S^{-\varepsilon}}, \varepsilon > 0\} \tag{1.17}$$

is unitarily isomorphic to the closure of all generalized functions $v \in V(T)$ with support in the domains $\overline{S^{-\varepsilon}}$, $\varepsilon > 0$, because

$$H_+(S^c)^\perp = \bigvee_{\varepsilon > 0} H[(S^c)^\varepsilon]^\perp.$$

As we saw earlier, when a biorthogonal function (1.3) is available, the space $V(T)$ contains $C_0^\infty(T)$ and each generalized function $v \in V(T)$ is the limit of a sequence $v_n \in C_0^\infty(T)$—see (1.13). One can ask, is every function $v \in V(T)$ with support $\operatorname{Supp} v \subseteq \overline{S}^{-\varepsilon}$, $\varepsilon > 0$, the limit of some sequence $v_n \in C_0^\infty(S)$?

In the ensuing discussion we will give conditions under which the answer to this question is yes for open domains $S \subseteq T$ which are bounded or have bounded complements in T. In several cases the answer is even yes for all domains $S \subseteq T$.

Denote by $V(S)$ the closure of all generalized functions $v \in V(T)$ with $\operatorname{Supp} v \subseteq \overline{S}^{-\varepsilon}$, $\varepsilon > 0$. An affirmative answer to our question implies that

$$V(S) = \overline{C_0^\infty(S)},$$

i.e., $V(S)$ coincides with the closure of $C_0^\infty(S)$ in the space $V(T)$ and this is equivalent to the equality

$$H^*(S) = H_+(S^c)^\perp. \tag{1.18}$$

This is because for generalized functions $v \in C_0^\infty(S)$ there correspond, by (1.12), variables $\eta(v) = Lv$ whose closure is $H^*(S)$, so that we have

$$H^*(S) \leftrightarrow V(S)$$

in the sense of the unitary correspondence (1.9); by (1.17) we have

$$V(S) \leftrightarrow H_+(S^c)^\perp.$$

Notice that equation (1.18) applied to an appropriate system of domains $S \subseteq T$ expresses the duality of the random fields (1.2) and (1.6) generated by the biorthogonal generalized random functions (1.1) and (1.3); duality was introduced in Chapter 2, §3.5, in connection with the Markov property.

The question of duality (1.18) will have particular importance for us in the case where the operator L which gives the biorthogonal function (1.3) is *local* in the sense that

$$(Lu, Lv) = 0 \tag{1.19}$$

for functions $u, v \in C_0^\infty(T)$ having disjoint supports. It is known that a continuous bilinear form (Lu, Lv), $u, v \in C_0^\infty(T)$, which satisfies condition (1.19) can be represented in the form

$$(Lu, Lv) = \int \sum_{k, j} b_{kj}(t) D^k u(t) \overline{D^j v(t)} \, dt, \tag{1.20}$$

where $b_{kj}(t)$ are locally integrable functions such that in every compact domain only a finite number of them are not zero.

We state some additional conditions under which the random fields (1.2) and (1.6) will be dual.

Lemma 1. *Suppose that the operation of multiplication by a function $w \in C_0^\infty(\mathbb{R}^d)$ is bounded in the space $V(T)$ with norm (1.14):*

$$\|wu\|_L \le C\|u\|_L, \qquad u \in C_0^\infty(T). \tag{1.21}$$

Then the duality condition (1.18) is satisfied for all domains $S \subseteq T$ which are bounded or have bounded complements in T.

PROOF. We look at an arbitrary element $v \in V(T)$ with Supp $v \subseteq \overline{S^{-\varepsilon}}$ and a sequence of functions $u_n \in C_0^\infty(T)$ converging to it; such a sequence exists because $C_0^\infty(T)$ is dense in $V(T)$. Let S be a bounded domain and let $w \in C_0^\infty(\mathbb{R}^d)$ be a non-negative function such that $w(t) = 1$ on $\overline{S^{-\varepsilon}}$ and $w(t) = 0$ for $(S^c)^\delta$, $\delta < \varepsilon$.

Let $v_n = w \cdot u_n$. Clearly, $v_n \in C_0^\infty(S)$. By (1.21)

$$\|v_n\|_L \le C\|u_n\|_L \le C_1,$$

since the converging sequence u_n, $n = 1, 2, \ldots$, is bounded. Moreover, for any $u \in C_0^\infty(T)$,

$$(u, v_n) = (uw, u_n) \to (uw, v) = (u, v),$$

since $uw(t) = u(t)$ for $t \in \overline{S^{-\varepsilon}}$. By virtue of this convergence we note that

$$\begin{aligned}(u, v_n) &= (Ju, \eta(v_n)) = \langle v(Ju), v_n \rangle \to \langle v(Ju), v \rangle \\ &= (Ju, \eta(v)) = (u, v), \qquad u \in C_0^\infty(T),\end{aligned}$$

where, by (1.9), the generalized functions $v(Ju)$ corresponding to the variables $Ju \in H(T)$, $u \in C_0^\infty(T)$, form a dense set in the space $V(T)$ since the set Ju, $u \in C_0^\infty(T)$, is dense in $H(T)$. Thus the sequence of generalized functions v_n, $n = 1, 2, \ldots$, which is bounded in $V(T)$, converges weakly to $v \in V(T)$ and v belongs to the closed linear span of the elements $v_n \in C_0^\infty(S)$, which is what we wanted to show. In the case of an unbounded domain S with bounded complement $S^c = T \backslash S$, the proof is similar. One looks at a function $w \in C_0^\infty(\mathbb{R}^d)$ such that $w = 0$ on $\overline{S^{-\varepsilon}}$ and $w = 1$ on $(S^c)^\delta$, $\delta < \varepsilon$, and sets $v_n = (1 - w)u_n = u_n - wu_n$. The sequence $v_n \in C_0^\infty(S)$ is bounded, since

$$\|v_n\|_L \le \|u_n\|_L + \|wu_n\|_L,$$

and converges weakly to an element v in the space $V(T)$.

We remark that the condition for duality is symmetric for the biorthogonal functions (1.1), (1.3) and it will be met if condition (1.21) holds with respect to the norm

$$\|u\|_J = \|Ju\|, \qquad u \in C_0^\infty(T). \tag{1.22}$$

Denote by $V_A(T)$ the completion of $C_0^\infty(T)$ with respect to the corresponding norm $\|u\|_A$, where $A = J, L$.

Let $T = \mathbb{R}^d$ and define the shift operator τ_s:

$$\tau_s u(t) = u(t + s), \qquad u \in C_0^\infty(\mathbb{R}^d). \qquad \square$$

Lemma 2. *Suppose that with respect to both norms* $\|u\|_A$, $A = J, L$, *the shift operators are bounded for sufficiently small s:*

$$\|\tau_s u\|_A \leq C\|u\|_A, \qquad u \in C_0^\infty(\mathbb{R}^d), \tag{1.23}$$

in a sufficiently small neighborhood $|s| < \delta_0$. *Then condition* (1.18) *holds for all domains* $S \subseteq T$ *which are bounded or have bounded complements in* T.

PROOF. Let $w(t) = w(-t)$, $t \in \mathbb{R}^d$, be a symmetric non-negative function in $C_0^\infty(\mathbb{R}^d)$. It is known that for any generalized function $v = (u, v)$, $u \in C_0^\infty(\mathbb{R}^d)$, the *convolution*

$$w * v(s) = \overline{(\tau_{-s}w, v)}, \qquad s \in \mathbb{R}^d,$$

is infinitely-differentiable and furthermore

$$(u, w * v) = \int u(s)(\tau_{-s}w, v)\, ds = (w * u, v), \qquad u \in C_0^\infty(\mathbb{R}^d),$$

where $w * u(t)$

$$w * u(t) = \int w(t - s)u(s)\, ds, \qquad t \in \mathbb{R}^d.$$

We take an "approximate identity" sequence of non-negative symmetric functions $w = w_n$:

$$\int w_n(t)\, dt = 1, \qquad \int_{|t| \geq \delta} w_n(t)\, dt \to 0 \quad \text{as } n \to \infty,$$

for any $\delta > 0$. We consider a domain $S \subseteq T$ and generalized function $v \in V_A(T)$ with support $\operatorname{Supp} v \subset T\backslash(S^c)^\varepsilon$. We choose functions $w_n \in C_0^\infty(\mathbb{R}^d)$ such that for all n the supports $\operatorname{Supp} w_n$ lie in small neighborhoods of zero, all contained in the neighborhood $|t| < \min(\delta_0, \varepsilon)$. We let

$$v_n(s) = w_n * v(s), \qquad s \in T,$$

(recalling that $T = \mathbb{R}^d$). For a bounded domain S we have $v_n \in C_0^\infty(S)$ because for $s \in S^c$ the support of the function $\tau_{-s}w_n(t)$, $t \in T$, lies in the domain $(S^c)^\varepsilon$ and does not intersect $\operatorname{Supp} v$, so that $(\tau_{-s}w_n, v) = 0$ in some neighborhood of the point s. We also have

$$(u, v_n) = (w_n * u, v) \to (u, v), \qquad u \in C_0^\infty(T),$$

by the $C_0^\infty(T)$ convergence of $w_n * u \to u$, i.e., all the derivatives converge,

$$D^k(w_n * u) = w_n * D^k u \to D^k u,$$

and the supports of all the functions $w_n * u$ are contained in some compact set. If we show that the sequence v_n, $n = 1, 2, \ldots$, is bounded, $\|v_n\|_A \leq C$, then as in Lemma 1 this will establish that it converges weakly to the generalized function $v \in V_A(T)$; as a result v must belong to the closed linear span of the generalized functions $v_n \in C_0^\infty(S)$, which is what we want to prove.

We know that the function $v \in V_A(T)$ is the limit of some sequence $u_m \in C_0^\infty(T)$, $m = 1, 2, \ldots$. We will show that the functions

$$w_n * u_m = \int w_n(s)[\tau_{-s} u_m] \, ds \in C_0^\infty(T)$$

converge weakly to $v_n \in V_A(T)$ as $m \to \infty$. We have (e.g., for $A = L$)

$$|(u, u_m - v)| = |(Ju, \eta(u_m - v))| \leq \|Ju\| \cdot \|u_m - v\|_A \to 0, \qquad u \in C_0^\infty(T),$$

so that

$$(u, w_n * u_m) = (w_n * u, u_m) \to (w_n * u, v) = (u, w_n * v), \qquad u \in C_0^\infty(T).$$

Using the fact that the function $A\tau_{-s} u_m$, $s \in \mathbb{R}^d$ (with $A = J, L$) is strongly continuous into the space $H(T)$ and taking the limit of the underlying "integrating sums," we get the equation

$$\int w_n(s)[A\tau_{-s} u_m] \, ds = A \int w_n(s)[\tau_{-s} u_m] \, ds = A(u_m * w_n).$$

From this, using (1.23) and the boundedness of the convergent sequence u_m, $m = 1, 2, \ldots$,

$$\|w_n * u_m\|_A = \|A(w_n * u_m)\| = \left\| \int w_n(s)[A\tau_{-s} u_m] \, ds \right\|$$

$$\leq \sup_{|s| < \delta_0} \|A\tau_{-s} u_m\| \int w_n(s) \, ds \leq C \|u_m\|_A \leq C_1.$$

This inequality gives the boundedness of the elements v_n as limits of the respective weakly convergent sequences of elements $w_n * u_m$, $m = 1, 2, \ldots$:

$$\|v_n\|_A \leq \sup_m \|v_n * u_m\| \leq C_1.$$

As we already observed, this allows us to conclude that the sequence v_n, $n = 1, 2, \ldots$, converges weakly to the generalized function $v \in V_A(T)$ which therefore lies in the closed linear span of the elements $v_n \in C_0^\infty(S)$. For the case $A = L$ this proves (1.18) is true for a bounded domain $S \subseteq T$. For the case $A = J$ this proves the "symmetric" equality

$$H(S) = H_+^*(S^c)^\perp,$$

interchanging the roles of the biorthogonal functions (1.1) and (1.3).

Now we look at a domain $S \subseteq T$ whose complement $S^c = T \backslash S$ is bounded. We showed that for any ε-neighborhood $(S^c)^\varepsilon$

$$H[(S^c)^\varepsilon] = H_+^* \overline{(S^{-\varepsilon})}^\perp, \qquad S^{-\varepsilon} = T \backslash (S^c)^\varepsilon,$$

from which it follows that

$$H_+(S^c) = \bigcap_{\varepsilon > 0} H[(S^c)^\varepsilon] = \bigcap_{\varepsilon > 0} H_+^* \overline{(S^{-\varepsilon})}^\perp = \left[\bigvee_{\varepsilon > 0} H_+^* \overline{(S^{-\varepsilon})} \right]^\perp = H^*(S)^\perp,$$

and by the same token we showed that (1.18) holds for all domains $S \subseteq T$ which are bounded or have bounded complements. This completes the proof of Lemma 2. □

We remark here that the duality condition given by (3.24) in Chapter 2 is not satisfied in every case for all domains $S \subseteq T$. An example is provided by the derivative of "white noise"

$$(u, \xi) = \int u'(t)\eta(dt), \qquad u \in C_0^\infty(T),$$

on the real line $T = \mathbb{R}$ with the standard orthogonal measure

$$E|\eta(dt)|^2 = dt;$$

this has a biorthogonal function—Brownian motion on $\mathbb{R}\backslash\{0\}$ (see §4.1 below)—but not on $\mathbb{R}$ and the duality condition (Ch. 2, (3.24)) fails for certain unbounded domains S having unbounded complements. In fact, it is not hard to check that for a finite interval $S \subseteq T$ the associated space $H(S)$— the closure of the variables

$$(u, \xi) = \int u'(t)\eta(dt), \qquad u \in C_0^\infty(S)$$

consists of all stochastic integrals

$$\eta = \int v(t)\eta(dt),$$

where the function $v(t)$ is square-integrable, has support contained in S and satisfies

$$\int v(t) \, dt = 0;$$

this last condition may be omitted when S is an infinite interval. From this it is evident that

$$H(S) = H_+(S).$$

From what has been said one can see that the spaces $H_+(-\infty, \varepsilon)$ and $H_+(-\varepsilon, \infty)$ contain the variable

$$\eta = \int_{-\varepsilon}^{\varepsilon} \eta(dt),$$

which is orthogonal to the space $H_+(-\varepsilon, \varepsilon) = H(-\varepsilon, \varepsilon)$; thus for the domains $S' = (-\infty, \varepsilon)$, $S'' = (-\varepsilon, \infty)$ we have

$$H_+(S') \cap H_+(S'') \neq H_+(S' \cap S'').$$

3. The Markov Property for Generalized Functions

We call a generalized random function (1.1) *Markov* if the random field it generates (1.2) is Markov. We will be interested for the most part in the Markov property in the wide sense with respect to all domains $S \subseteq T$ which are bounded or have bounded complements in T.

As we have seen, if the duality condition (1.18) holds, the Markov property for the random field (1.2) is equivalent to the statement that the dual field (1.6) is orthogonal. See Chapter 2, §3.5.

It is obvious that the orthogonality condition,

$$H^*(S') \perp H^*(S'') \tag{1.24}$$

for disjoint domains S', S'', will be satisfied if the generalized function (1.3) generating the random field (1.6) is given by a local operator L. Of course, the converse statement is also true, namely, if the orthogonality condition (1.24) holds then the operator L is local,

$$(Lu, Lv) = 0 \tag{1.25}$$

for any $u, v \in C_0^\infty(T)$ with disjoint supports; these could be defined on disjoint domains $S' \supseteq \mathrm{Supp}\, u$, $S'' \supseteq \mathrm{Supp}\, v$ of the type we are considering.

We call a generalized random function

$$(u, \zeta^*) = Lu, \qquad u \in C_0^\infty(T),$$

local if the operator L is local (it is also called a *generalized random function with orthogonal values*). We call the biorthogonal functions (1.1) and (1.3) *dual* if the duality condition (1.18) holds for them. Keeping the equivalence of conditions (1.24) and (1.25) in mind, we get from the theorem in §3.5, Chapter 2, the following criterion for the Markov behavior of dual random functions (1.1), (1.3).

Theorem. *A generalized random function is Markov if and only if its dual function is local.*

We should specify that the theorem refers to the Markov property with respect to the same system of domains $S \subseteq T$ with boundaries $\Gamma = \partial S$ separating $S_1 = S$ and $S_2 = T \backslash \bar{S}$ for which the conjugacy holds. This could be the system of all open domains $S \subseteq T$ which are bounded or have bounded complements in T, or it could be the system of all open domains $S \subseteq T$. Recall that some conditions for duality were given in Lemmas 1 and 2.

We note that the local property for the biorthogonal function (1.3) is always a necessary condition for the function (1.1) to be Markov. This is because for any functions $u, v \in C_0^\infty(T)$ with disjoint supports one can find an underlying domain $S \subseteq T$ such that

$$\mathrm{Supp}\, u \subseteq S_1 = S, \qquad \mathrm{Supp}\, v \subseteq S_2 = T \backslash \bar{S};$$

thus by the general inclusion (1.16) the variables

$$Lu \in H_+(S_1^c)^\perp, \qquad Lv \in H_+(S_2^c)^\perp$$

are orthogonal if the random field (1.2) is Markov since the Markov property gives (cf. (3.14) in Chapter 2)

$$H_+(S_1^c)^\perp \perp H_+(S_2^c)^\perp.$$

Earlier we gave a criterion for the Markov property for generalized random functions. Obviously it also applies to "ordinary" random functions

$$\xi(t) \in L^2(\Omega, \mathcal{A}, P), \qquad t \in T. \tag{1.26}$$

We call the random function (1.26) *Markov* if the random field it generates

$$H(S) = \bigvee_{t \in S} \xi(t), \qquad S \subseteq T, \tag{1.27}$$

is Markov. If the function $\xi(t)$ is weakly continuous then for any open domain $S \subseteq T$ the space $H(S)$ coincides with the closure of the variables

$$(u, \xi) = \int u(t)\xi(t)\, dt, \qquad u \in C_0^\infty(S), \tag{1.28}$$

so that the random function $\xi(t), t \in T$, is Markov if and only if the generalized function (1.28) also has this property. Notice here that we have specified the *generalized* random function $\xi(t), t \in T$, taking this to be

$$(u, \xi) = \int u(t)\overline{\xi(t)}\, dt, \qquad u \in C_0^\infty(T). \tag{1.29}$$

The random fields generated by the generalized functions (1.28) and (1.29) can be obtained from one another by a unitary transformation U on the space $L^2(\Omega, \mathcal{A}, P)$ by taking the complex conjugates of the variables. Clearly, for any unitary operator U the mapping

$$H(S) \to UH(S), \qquad S \subseteq T, \tag{1.30}$$

preserves the Markov property so that we have:

A weakly continuous random function $\xi(t), t \in T$, is Markov if and only if the related generalized random function (1.29) is Markov.

§2. Stationary Generalized Functions

1. Spectral Representation of Coupled Stationary Generalized Functions

A generalized random function

$$(u, \xi) = Ju, \qquad u \in C_0^\infty(\mathbb{R}^d) \tag{2.1}$$

is called *stationary* if the *shift operator*

$$U_s(u, \xi) = (\tau_{-s}u, \xi), \qquad s \in \mathbb{R}^d, \tag{2.2}$$

where $\tau_s u(t) = u(t + s)$, $t \in \mathbb{R}^d$, is isometric on $u \in C_0^\infty(\mathbb{R}^d)$ and extends to a unitary operator on the space $H(\mathbb{R}^d)$, the closure of the variables

$$(u, \xi) \in L^2(\Omega, \mathscr{A}, P), \qquad u \in C_0^\infty(\mathbb{R}^d).$$

It is well known that a stationary function admits a *spectral representation*

$$(u, \xi) = \int \tilde{u}(\lambda)\Phi(d\lambda), \qquad u \in C_0^\infty(\mathbb{R}^d) \tag{2.3}$$

where

$$\tilde{u}(\lambda) = \int \exp(i\lambda t)u(t)\, dt, \qquad \lambda \in \mathbb{R}^d,$$

is the (inverse) Fourier transform† and $\Phi(d\lambda)$, $\lambda \in \mathbb{R}^d$, is a stochastic orthogonal measure,

$$E|\Phi(d\lambda)|^2 = F(d\lambda),$$

satisfying the condition

$$\int (1 + |\lambda|^2)^{-n} F(d\lambda) < \infty \tag{2.4}$$

for some $n > 0$; the existence of such n is necessary and sufficient for the continuity of the functional

$$\int |\tilde{u}(\lambda)|^2 F(d\lambda) = \|(u, \xi)\|^2, \qquad u \in C_0^\infty(\mathbb{R}^d),$$

on the space $C_0^\infty(\mathbb{R}^d)$. F is usually called the *spectral measure* of the stationary function (2.1).

We look at the Hilbert space of measurable functions $\varphi(\lambda)$, $\lambda \in \mathbb{R}^d$, which are square-integrable with respect to the spectral measure $F(d\lambda)$; denote this space simply by $\mathscr{L}^2$. Its scalar product is

$$(\varphi, \psi) = \int \varphi(\lambda)\overline{\psi(\lambda)}F(d\lambda), \qquad \lambda, \psi \in \mathscr{L}^2.$$

We will show that the functions

$$\varphi(\lambda) = \tilde{u}(\lambda), \qquad u \in C_0^\infty(\mathbb{R}^d),$$

form a complete system in this space. To begin with, notice that one can approximate any function $\varphi(\lambda) \in C_0^\infty(\mathbb{R}^d)$ by functions $\tilde{u}(\lambda)$. In fact, all the derivatives $D^k v(t)$ of the Fourier transform

$$v(t) = \frac{1}{(2\pi)^d} \int \exp(-i\lambda t)\overline{\varphi(\lambda)}\, d\lambda, \qquad t \in R^d,$$

---

† Here and elsewhere we shall use the term "Fourier transform" to refer to the usual Fourier transform or its inverse.

are integrable and can be approximated arbitrarily closely by the corresponding derivatives $D^k u(t)$, $u \in C_0^\infty(\mathbb{R}^d)$, in the metric

$$\int |D^k v(t) - D^k u(t)| \, dt.$$

For instance, for the approximating functions one can use

$$u(t) = v(t)w(t/\tau), \qquad \tau \to \infty,$$

where the function $w(t) \in C_0^\infty(\mathbb{R}^d)$ is equal to 1 on $|t| \leq 1$; this works because $D^k u(t) = D^k v(t)$ for $|t| < \tau$ and

$$|D^k u(t)| \leq C \sum_{|j| \leq |k|} |D^j v(t)|, \qquad |t| \geq \tau,$$

where C is a constant depending on the derivatives $D^j w$, $|j| \leq |k|$. If we use derivatives D^k, $|k| \leq 2n$, to form the differential operator $(1 - \Delta)^n$, where Δ is the Laplacian operator, we get

$$(1 + |\lambda|^2)^l |\varphi(\lambda) - \tilde{u}(\lambda)| \leq \int |(1 - \Delta)^l v(t) - (1 - \Delta^l)u(t)| \, dt \to 0.$$

From this we get that for the spectral measure $F(d\lambda)$ satisfying (2.4) and for $2l \geq n$,

$$\int |\varphi(\lambda) - \tilde{u}(\lambda)|^2 F(d\lambda) \leq \sup_\lambda (1 + |\lambda|^2)^{2l} |\varphi(\lambda) - \tilde{u}(\lambda)|^2 \int (1 + |\lambda|^2)^{-2l} F(d\lambda) \to 0.$$

Furthermore, every continuous function $\varphi(\lambda)$ with compact support Supp φ can be uniformly approximated by functions from $C_0^\infty(\mathbb{R}^d)$ with supports in arbitrarily small ε-neighborhoods of the set Supp φ (by taking the convolution of φ with a standard approximate identity sequence, for example). Finally, these continuous functions $\varphi(\lambda)$ can be approximated with indicator functions and these form a complete system in our space $\mathscr{L}^2$.

From the completeness of the set of functions $\tilde{u}(\lambda)$ arising in the spectral representation (2.3) it follows that the space $H(\mathbb{R}^d)$ is the collection of all variables of the form

$$\eta = \int \varphi(\lambda)\Phi(d\lambda), \qquad \int |\varphi(\lambda)|^2 F(d\lambda) < \infty, \tag{2.5}$$

and moreover is unitarily isomorphic to $\mathscr{L}^2$:

$$(\eta_1, \eta_2) = \int \varphi_1(\lambda)\overline{\varphi_2(\lambda)} \, F(d\lambda)$$

for any $\eta_1, \eta_2 \in H(\mathbb{R}^d)$ and functions $\varphi_1, \varphi_2 \in \mathscr{L}^2$ associated with them by formula (2.5).

The stochastic orthogonal measure $\Phi(d\lambda)$ in the spectral representation (2.3) is closely related to the usual spectral representation

$$U_t = \int \exp(i\lambda t) E(d\lambda), \quad t \in \mathbb{R}^d, \tag{2.6}$$

of the group of unitary shift operators, where $E(d\lambda)$, $\lambda \in \mathbb{R}^d$, is a spectral family of projection operators on the space $H(\mathbb{R}^d)$. To elaborate, for every variable $\eta \in H(\mathbb{R}^d)$ the equation

$$\Phi_\eta(\Delta) = E(\Delta)\eta, \quad \Delta \subseteq \mathbb{R}^d, \tag{2.7}$$

defines an orthogonal measure Φ_η on Borel sets $\Delta \subseteq \mathbb{R}^d$, and for $\eta = (u, \xi)$ we have

$$U_t \eta = \int \exp(i\lambda t)\tilde{u}(\lambda)\Phi(d\lambda) = \int \exp(i\lambda t)\Phi_\eta(d\lambda), \quad t \in \mathbb{R}^d,$$

from which it follows that

$$\tilde{u}(\lambda)\Phi(d\lambda) = \Phi_\eta(d\lambda), \quad u \in C_0^\infty(\mathbb{R}^d). \tag{2.8}$$

We use equation (2.8) to conclude that

$$E(\Delta) \int \varphi(\lambda)\Phi(d\lambda) = \int_\Delta \varphi(\lambda)\Phi(d\lambda) \tag{2.9}$$

for any Borel $\Delta \subseteq \mathbb{R}^d$ and $\varphi \in \mathcal{L}^2$.

We remark that for a fixed function $\varphi \in \mathcal{L}^2$ the system of variables

$$\int_\Delta \varphi(\lambda)\Phi(d\lambda), \quad \Delta \subseteq \mathbb{R}^d, \tag{2.10}$$

is complete in the space $H(\mathbb{R}^d)$ if and only if $\varphi(\lambda) \neq 0$ almost everywhere with respect to the measure F.

We say that the stationary generalized random function

$$(u, \xi^*) = Lu, \quad u \in C_0^\infty(\mathbb{R}^d), \tag{2.11}$$

with values $(u, \xi^*) \in H(\mathbb{R}^d)$ is *coupled* with the stationary generalized function (2.1) if

$$(\tau_{-s}u, \xi^*) = U_s(u, \xi^*), \quad s \in \mathbb{R}^d, \tag{2.12}$$

for all $u \in C_0^\infty(\mathbb{R}^d)$, where U_s is the unitary shift operator on the space $H(\mathbb{R}^d)$ defined by (2.2).

Let us look at the spectral representation of the stationary function (2.11):

$$(u, \xi^*) = \int \tilde{u}(\lambda)\Phi^*(d\lambda), \quad u \in C_0^\infty(\mathbb{R}^d). \tag{2.13}$$

When this is coupled with the stationary function (2.1) it has the same group of unitary shift operators (2.6) and the space $H^*/\mathbb{R}^d)$—the closure of all variables (2.11)—is invariant with respect to these. In this case, the space $H^*(\mathbb{R}^d)$ is also invariant under the family of projections $E(\Delta)$, $\Delta \subseteq \mathbb{R}^d$, and on the basis of relations similar to (2.7), (2.8) one arrives at the same results for $\Phi^*(d\lambda)$ as for $\Phi(d\lambda)$ (cf. (2.9) and (2.10)). In particular, taking the variable $\eta = (u, \xi^*)$,

$$\eta = \int \tilde{u}(\lambda)\Phi^*(d\lambda) = \int \varphi(\lambda)\Phi(d\lambda),$$

we get, according to (2.9), that

$$E(\Delta)\eta = \int_\Delta \tilde{u}(\lambda)\Phi^*(d\lambda) = \int_\Delta \varphi(\lambda)\Phi(d\lambda), \qquad \Delta \subseteq \mathbb{R}^d.$$

Taking the variable $\eta = (u, \xi^*)$ for a function $u \in C_0^\infty(\mathbb{R}^d)$ whose Fourier transform $\tilde{u}(\lambda)$, $\lambda \in \mathbb{R}^d$, does not vanish, we see that

$$\Phi^*(d\lambda) = \gamma(\lambda)\Phi(d\lambda), \tag{2.14}$$

where $\gamma(\lambda) = \varphi(\lambda)/\tilde{u}(\lambda)$. From this,

$$H^*(\mathbb{R}^d) = H(\mathbb{R}^d) \tag{2.15}$$

if and only if $\gamma(\lambda) \neq 0$ almost everywhere with respect to the spectral measure $F(d\lambda)$, $\lambda \in \mathbb{R}^d$, since the space $H^*(\mathbb{R}^d)$ coincides with the closed linear span of the variables

$$\int_\Delta \tilde{u}(\lambda)\Phi^*(d\lambda) = \int_\Delta \tilde{u}(\lambda)\gamma(\lambda)\Phi(d\lambda), \qquad \Delta \subseteq \mathbb{R}^d.$$

2. Biorthogonal Stationary Functions

Suppose the stationary random function (2.1) has a biorthogonal function

$$(u, \xi^*) = Lu, \qquad u \in C_0^\infty(\mathbb{R}^d).$$

The biorthogonality condition (1.4) implies that

$$(Ju, U_s Lv) = (U_{-s}Ju, Lv) = (J\tau_s u, Lv) = (\tau_s u, v)$$
$$= (u, \tau_{-s}v) = (Ju, L\tau_{-s}v)$$

for all $u, v \in C_0^\infty(\mathbb{R}^d)$ and $s \in \mathbb{R}^d$, hence

$$U_s Lv = L\tau_{-s}v$$

and equation (2.12) holds for the unitary operators (2.2). We have shown:

A generalized random function which is biorthogonal to a stationary function (2.1) is coupled with it.

We look at the spectral representation (2.13) of the biorthogonal stationary function (2.11), in which the orthogonal measure $\Phi^*(d\lambda)$ is given by (2.14) with "density" $\gamma(\lambda)$, $\lambda \in \mathbb{R}^d$, as we have just seen. The function $\gamma(\lambda) \neq 0$ almost everywhere with respect to the spectral measure $F(d\lambda)$ because by (1.5) we have equation (2.15) holding for the biorthogonal function. Moreover the function $\gamma(\lambda)$ can be defined from the biorthogonality condition (1.4):

$$(Ju, Lv) = \int \tilde{u}(\lambda)\overline{\tilde{v}(\lambda)}\overline{\gamma(\lambda)}F(d\lambda) = \int u(t)\overline{v(t)}\, dt$$

$$= \frac{1}{(2\pi)^d} \int \tilde{u}(\lambda)\overline{\tilde{v}(\lambda)}\, d\lambda, \qquad u, v \in C_0^\infty(\mathbb{R}^d),$$

giving

$$\overline{\gamma(\lambda)}F(d\lambda) = \frac{1}{(2\pi)^d}\, d\lambda, \qquad \gamma(\lambda) \geq 0.$$

It is evident that there exists a *spectral density*

$$f(\lambda) = \frac{F(d\lambda)}{d\lambda} = \frac{1}{(2\pi)^d\gamma(\lambda)}$$

and

$$\gamma(\lambda) = \frac{1}{(2\pi)^d}\, f(\lambda)^{-1}; \tag{2.16}$$

recall that $\gamma(\lambda) > 0$ a.e. with respect to $F(d\lambda)$. From this, using (2.14), the spectral density of the biorthogonal stationary function will be

$$f^*(\lambda) = |\gamma(\lambda)|^2 f(\lambda) = \frac{1}{(2\pi)^{2d}}\, f(\lambda)^{-1}. \tag{2.17}$$

Like every spectral density of a stationary generalized function, $f^*(\lambda)$ must satisfy a condition of the type (2.4), namely

$$\int (1 + |\lambda|^2)^{-m} f^*(\lambda)\, d\lambda < \infty \tag{2.18}$$

for some $m > 0$. With (2.17) in mind, we conclude that *the condition*

$$\int (1 + |\lambda|^2)^{-m} f(\lambda)^{-1}\, d\lambda < \infty \tag{2.19}$$

for some $m > 0$ is necessary and sufficient for the existence of the biorthogonal function.

When this condition is satisfied, the biorthogonal function (2.11) is defined by means of the spectral representation (2.13), in which the orthogonal measure $\Phi^*(d\lambda)$, $\lambda \in \mathbb{R}^d$, is given by formula (2.14) with a density $\gamma(\lambda)$ of the form (2.16).

In concluding this section it seems relevant to the topic of stationary generalized functions to make a general remark about the biorthogonality conditions (1.4), (1.5). Suppose the stationary function (2.1) has a spectral measure $F(d\lambda)$ which is not absolutely continuous but whose absolutely continuous component has a density $f(\lambda)$ of the type (2.19). Then the generalized function (2.11) which is coupled with (2.1) and has spectral density (2.17) satisfies condition (1.4) but does not satisfy condition (1.5)—for this function $H^*(\mathbb{R}^d) \subset H(\mathbb{R}^d)$.

3. The Duality Condition and a Markov Criterion

A biorthogonal stationary function is always dual with respect to the system of open domains $S \subseteq \mathbb{R}^d$ which are bounded or have bounded complements in $\mathbb{R}^d$. Indeed, condition (1.23) is satisfied by the operators $A = J, L$ because

$$\|\tau_s u\|_A = \|A\tau_s u\| = \|U_{-s}Au\| = \|Au\| = \|u\|_A, \qquad s \in \mathbb{R}^d,$$

for all $u \in C_0^\infty(\mathbb{R}^d)$.

Thus, by the theorem in §1, one of a pair of biorthogonal stationary functions is Markov if and only if the other is local.

We will examine the question of Markov behavior of a stationary generalized function (2.1). We know that it is necessary and sufficient for the existence of a biorthogonal function (2.11) that the stationary function (2.1) have a spectral density $f(\lambda)$ satisfying (2.19); from this,

$$(Lu, Lv) = \int \tilde{u}(\lambda)\overline{\tilde{v}(\lambda)} f^*(\lambda)\, d\lambda, \qquad u, v \in C_0^\infty(\mathbb{R}^d), \tag{2.20}$$

where the spectral density $f^*(\lambda)$ of the stationary function (2.11) is defined in (2.17). One can ask what the function $f^*(\lambda)$ must be like in order for the bilinear form (2.20) to vanish for all $u, v \in C_0^\infty(\mathbb{R}^d)$ with disjoint supports.

Notice that the product $\tilde{u}(\lambda) \cdot \overline{\tilde{v}(\lambda)}$ is the Fourier transform of the convolution

$$v * u(t) = \overline{(\tau_{-t}v, u)} = \int \overline{v(s - t)}u(s)\, ds, \qquad t \in \mathbb{R}^d,$$

so that on the right-hand side of (2.20) we are concerned with values of the generalized function

$$B^*(u) = \int \tilde{u}(\lambda)f^*(\lambda)\, d\lambda, \qquad u \in C_0^\infty(\mathbb{R}^d), \tag{2.21}$$

(recall here that the spectral density $f^*(\lambda)$ satisfies (2.18)). Clearly the operator L will satisfy the local property (see (1.25)) implicit in our question if the generalized function (2.21) has support equal to the set {0}. This is because for functions $u, v \in C_0^\infty(\mathbb{R}^d)$ with disjoint supports

$$v * u(t) = \int \overline{v(s - t)} u(s) \, ds = 0$$

in some neighborhood $|t| < \varepsilon$; on the other hand, taking an arbitrary function $u \in C_0^\infty(\mathbb{R}^d)$ with $0 \notin \text{Supp } u$ and an approximate identity sequence of symmetric non-negative functions $v = w_n \in C_0^\infty(\mathbb{R}^d)$ with supports in small neighborhoods of zero (disjoint from Supp u), we get from the local behavior of the operator L (see (1.25)) that

$$B^*(u) = \lim_{n \to \infty} B^*(w_n * u) = \lim_{n \to \infty} (Lu, Lw_n) = 0$$

and thus

$$\text{Supp } B^* = \{0\}. \tag{2.22}$$

It is well known that an arbitrary generalized function $B^*(u)$, $u \in C_0^\infty(\mathbb{R}^d)$, whose support is {0} is a linear combination of the "δ-function" and its derivatives,

$$B^*(u) = \sum_{|k| \leq 2l} b_k D^k u(0), \qquad u \in C_0^\infty(\mathbb{R}^d),$$

and for the function (2.21) this is equivalent to its generalized Fourier transform $f^*(\lambda)$, $\lambda \in \mathbb{R}^d$, being a polynomial:

$$f^*(\lambda) = \frac{1}{(2\pi)^d} \sum_{|k| \leq 2l} b_k (-i\lambda)^k, \tag{2.23}$$

where we have, using coordinate notation $\lambda = (\lambda_1, \ldots, \lambda_d)$

$$(-i\lambda)^k = (-i\lambda_1)^{k_1} \cdots (-i\lambda_d)^{k_d} \quad \text{for } k = (k_1, \ldots, k_d).$$

One can also say that the covariance functional (2.20) is given by a differential form

$$(Lu, Lv) = (\mathscr{P}u, v), \qquad u, v \in C_0^\infty(\mathbb{R}^d), \tag{2.24}$$

with

$$\mathscr{P} = \sum_{|k| \leq 2l} b_k D^k$$

a positive linear differential operator with constant coefficients b_k, $|k| \leq 2\lambda$.

In other words, taking formula (2.17) into account, we have the following result.

Theorem. *A stationary generalized function with spectral density $f(\lambda)$ satisfying (2.19) is Markov if and only if the function $1/f(\lambda)$ is a polynomial.*

We stress that here we are speaking of the Markov property with respect to the system of open domains $S \subseteq \mathbb{R}^d$ which are bounded or have bounded complements in $\mathbb{R}^d$, with boundaries $\Gamma = \partial S$ separating the domains $S_1 = S$ and $S_2 = \mathbb{R}^d \backslash \bar{S}$.

We remark in connection with the theorem above that the Markov property, as we are considering it, is well known for stationary processes (i.e., stationary random functions $z(t)$, $-\infty < t < \infty$) with spectral densities of the form

$$f(\lambda) = \frac{1}{2\pi} \frac{1}{|L(i\lambda)|^2}, \qquad -\infty < 1 < \infty,$$

where

$$L(z) = \sum_k a_k z^k$$

is a polynomial, all of whose zeros lie in the left half-plane $\text{Re } z < 0$. In particular, $\xi(t)$ is the stationary solution of the stochastic differential equation considered earlier in Chapter 2 (1.21),

$$L\xi(t) = \sum_{k=0}^{l} a_k \frac{d^k}{dt^k} \xi(t) = \dot{\eta}(t), \qquad (2.25)$$

with constant coefficients a_k, $0 \le k \le l$, and standard "white noise"

$$(u, \dot{\eta}) = \int u(t)\eta(dt), \qquad u \in C_0^\infty(-\infty, \infty),$$

on the right side. For each t, the white noise on the half-line $(-\infty, t)$ introduced here generates the "past," $H(-\infty, t)$, of the process under consideration; to be precise, the space $H(-\infty, t)$ is the closure of all variables of the form

$$\eta = \int u(s)\eta(ds), \qquad u \in C(-\infty, t).$$

The stationary solution of equation (2.25) can be obtained, for example, by taking the limit as $t_0 \to -\infty$ of the solutions of this equation on the half-line (t_0, ∞), with given initial conditions $\xi(t_0)^{(k)}$, $0 \le k \le l - 1$; this differs from the solution with initial conditions $\xi(t_0)^{(k)} = 0$, $0 \le k \le l - 1$, considered in Chapter 2, §1, only in a term which decays exponentially as $t_0 \to -\infty$ and which solves the homogeneous equation (2.25) with right side zero and given "stationary" initial conditions $\xi(t_0)^{(k)}$, $0 \le k \le l - 1$. As we essentially showed in Chapter 2, the random function $\xi(t)$, $-\infty < t < \infty$, together with its derivatives, forms an l-dimensional Markov process in the traditional sense,

$$(\xi^{(k)}(t), 0 \le k \le l - 1\}, \qquad -\infty < t < \infty.$$

§3. Biorthogonal Generalized Functions Given by a Differential Form

1. Basic Definitions

It is known that for each continuous bilinear form

$$B^*(u, v), \qquad u, v \in C_0^\infty(T), \tag{3.1}$$

T a domain $\subseteq \mathbb{R}^d$, which is positive in the sense that

$$B^*(u, u) \geq 0, \qquad u \in C_0^\infty(T),$$

there exists a generalized random function

$$(u, \xi^*) = Lu, \qquad u \in C_0^\infty(T), \tag{3.2}$$

having (3.1) as its covariance function:

$$(Lu, Lv) = B^*(u, v), \qquad u, v \in C_0^\infty(T).$$

We specify that the operator L giving the generalized function (3.2) be regarded as a linear operator

$$L: C_0^\infty(T) \subseteq L^2(T) \to H^*(T),$$

mapping its domain $C_0^\infty(T) \subseteq L^2(T)$ into the space $H^*(T)$. Here $L^2(T)$ is the Hilbert space of square-integrable functions $u(t)$, $t \in T$, with norm

$$\|u\| = \left(\int_T |u(t)|^2 \, dt \right)^{1/2},$$

and $H^*(T)$ is the closure of the variables

$$Lu \in L^2(\Omega, \mathscr{A}, P), \qquad u \in C_0^\infty(T).$$

Let (3.1) be a differential form of the following sort:

$$B^*(u, v) = \sum_{k,j} \int b_{kj} D^k u(t) \overline{D^j v(t)} \, dt, \qquad u, v \in C_0^\infty(T), \tag{3.3}$$

with coefficients $b_{kj}(t)$ which are sufficiently smooth and bounded; the index $k = (k_1, \ldots, k_d)$ is a vector with positive integer coordinates, $|k| = k_1 + \cdots + k_d$, and

$$D^k = \frac{\partial^{|k|}}{\partial t_1^{k_1} \cdots \partial t_d^{k_d}}$$

is the usual partial differentiation operator in the open domain $T \subseteq \mathbb{R}^d$. We will be concerned with a differential form of order $2l$, with only a finite number of coefficients $b_{kj}(t)$, $|k| + |j| \leq 2l$, different from zero. Such a differential form can be integrated by parts, leading to a version with the same type of coefficients but with $b_{kj}(t)$ different from zero only when $|k|, |j| \leq l$:

$$B^*(u, v) = \int \sum_{|k|, |j| \leq l} b_{kj}(t) D^k u(t) \overline{D^j v(t)}\, dt;$$

as an example of this "index symmetrization," notice that a form $B^*(u, v)$ is positive definite if for all $t \in T$ the corresponding quadratic form

$$\sum_{k, j} b_{kj}(t) x_k \overline{x}_j \geq 0$$

for all scalar variables x_k, $|k| \leq l$.

Again using integration by parts, we can represent the differential form (3.3) in the following way:

$$B^*(u, v) = (\mathscr{P}u, v), \qquad u, v \in C_0^\infty(T), \tag{3.4}$$

where

$$\mathscr{P} = \sum_{|k| \leq 2l} b_k(t) D^k$$

is a linear differential operator; positive definiteness of the differential form (3.4) means that the associated differential operator $\mathscr{P}$ is positive:

$$(\mathscr{P}u, u) \geq 0, \qquad u \in C_0^\infty(T).$$

We have already encountered an example of a differential form of type (3.4) which is important for us: see (2.24) in our discussion of stationary generalized functions.

Let a differential form (3.3) be *nondegenerate*, in the sense that

$$\|Lu\|^2 = B^*(u, u) \geq c\|u\|^2, \qquad u \in C_0^\infty(T), \tag{3.5}$$

for some constant $c > 0$.

For example, notice that this condition will be fulfilled if the quadratic form for the symmetrized version of (3.3), with coefficients $b_{kj}(t)$, $|k|, |j| \leq l$, is nondegenerate in the variable x_0, i.e.

$$\sum_{k, j} b_{kj}(t) x_k \overline{x}_j \geq c|x_0|^2, \qquad t \in T,$$

for all x_k, $|k| \leq l$.

Notice also that for a differential form given as in (3.4) by a positive, linear differential operator $\mathscr{P}$, condition (3.5) takes the form

$$(\mathscr{P}u, u) \geq c\|u\|^2, \qquad u \in C_0^\infty(T),$$

and this is known to be equivalent to the operator $\mathscr{P}$ having a self-adjoint extension with a bounded inverse operator $\mathscr{P}^{-1}$. In particular, condition (3.5) holds when

$$\mathscr{P} = \sum_k b_k D^k$$

is an operator with constant coefficients for which the corresponding polynomial

$$\mathscr{P}(-i\lambda) = \sum_k b_k \cdot (-i\lambda)^k, \qquad \lambda \in \mathbb{R}^d,$$

is strictly positive; using the Fourier transform $\tilde{u}(\lambda)$ of a function $u \in C_0^\infty(T)$ we have

$$\|Lu\|^2 = \frac{1}{(2\pi)^d} \int |\tilde{u}(\lambda)|^2 \mathscr{P}(-i\lambda)\, d\lambda \geq c\|u\|^2, \qquad (3.6)$$

if $\mathscr{P}(-i\lambda) \geq c > 0$.

The *closure* of the operator L—also denoted L— is defined by the equation

$$Lu = \eta \qquad (3.7)$$

for all $u \in L^2(T)$ for which there exists a sequence $u_n \in C_0^\infty(T)$ such that

$$Lu_n \to \eta, \qquad u_n \to u.$$

For every variable $\eta \in H^*(T)$ one can find a sequence $u_n \in C_0^\infty(T)$ such that $Lu_n \to \eta$, and by (3.5) we have

$$\|u_n - u_m\| \leq C\|Lu_n - Lu_m\| \to 0, \qquad n, m \to \infty.$$

Thus the variable η belongs to the range of the closure L: $\eta = Lu$, where the function $u \in L^2(T)$ is the limit of an $L^2(T)$-Cauchy sequence $u_n \in C_0^\infty(T)$.

Denote by $D(t)$ the domain of definition of the closed operator (3.7). It is clear that there exists a bounded inverse

$$L^{-1}: H^*(T) \to D(T) \supseteq C_0^\infty(T).$$

The adjoint operator

$$J = (L^{-1})^*: L^2(T) \to H^*(T)$$

is defined on the whole space $L^2(T)$, is bounded, and has a range space which is dense in $H^*(T)$, since its adjoint operator $J^* = L^{-1}$ is nondegenerate. Clearly the generalized function

$$(u, \xi) = Ju, \qquad u \in C_0^\infty(T), \qquad (3.8)$$

is biorthogonal to the generalized function (3.2):

$$(Ju, Lv) = (u, J^*Lv) = (u, v), \qquad u, v \in C_0^\infty(T)$$

and the closure $H(T)$ of the variables $Ju, u \in C_0^\infty(T)$, coincides with the space $H^*(T)$ because for the bounded operator J this closure coincides with the closure of the range $Ju, u \in L^2(T)$.

We turn our attention to the reproducing kernel space $V(T)$ for the generalized function (3.8). It can be identified with the domain $D(T) \subseteq L^2(T)$ of the closed operator (3.7). Namely, each variable $\eta \in H(T)$ can be represented in the form

$$\eta = Lv, \qquad v \in D(T), \qquad (3.9)$$

and its associated generalized function $v(\eta) \in V(T)$ is

$$(u, v(\eta)) = (Ju, Lv) = (u, J^*Lv) = (u, v) = \int u(t)\overline{v(t)}\, dt, \qquad u \in C_0^\infty(T),$$

i.e., $v(\eta)$ coincides, as a generalized function, with the function $v \in D(T)$. Recall that the scalar product on the Hilbert space

$$V(T) = D(T) \tag{3.10}$$

is given by (cf. (1.8), (1.10))

$$\langle u, v \rangle = (Lu, Lv), \qquad u, v \in D(T). \tag{3.11}$$

2. Conditions for Markov Behavior

We will consider the question of the duality of the biorthogonal generalized functions (3.2) and (3.8) which correspond to a differential form (3.3) satisfying condition (3.5).

Let $S \subseteq T$ be an open domain with boundary ∂S in $\mathbb{R}^d$ which coincides with its boundary $\Gamma = \partial S$ in the domain T. (Obviously, when $T = \mathbb{R}^d$ this imposes no restrictions on the domain $S \subseteq \mathbb{R}^d$.)

We look at a function $v \in D(T)$ in the domain of the closed operator (3.7) with support $\operatorname{Supp} v \subseteq S^{-\varepsilon}$ for some $\varepsilon > 0$; recall that $S^{-\varepsilon} = T \setminus (S^c)^\varepsilon$ and $(S^c)^\varepsilon$ is an ε-neighborhood of S^c, the complement of the domain $S \subseteq T$. We require that $\operatorname{Supp} v$ lie a positive distance $\varepsilon > 0$ from the boundary of the domain S. We suppose that for each such function $v \in D(T)$ the shift

$$\tau_s v(t) = v(t + s), \qquad t \in T,$$

gives, for small s, $|s| < \delta$, a function in the space $D(T)$ and in addition

$$\|\tau_s v\|_L \leq C, \tag{3.12}$$

where the constants δ and C depend on v.

We will show that the function $v \in D(T)$ is the weak limit in the Hilbert space $V(T) = D(T)$ (cf. (3.10)) of some sequence $u_n \in C_0^\infty(S)$; this will mean that the duality condition (1.18) is true for the domain $S \subseteq T$.

It is clear that the differential form (3.3), together with the operator L, is continuous with respect to the norm

$$\|u\|_l = \left(\sum_{|k| \leq l} \|D^k u\|^2 \right)^{1/2}; \tag{3.13}$$

namely, making use of the symmetric version of a differential form of order $2l$ with bounded coefficients $b_{kj}(t), |k|, |j| \leq l$, we arrive at the obvious inequality

$$\|Lu\|^2 \leq C \sum_{|k| \leq l} \|D^k u\|^2, \qquad u \in C_0^\infty(T).$$

Thus, instead of functions $u_n \in C_0^\infty(S)$ approximating $v \in D(T)$ we can take functions $u_n \in \mathring{W}_2^l(S)$, where $\mathring{W}_2^l(S)$ is the Sobolev space consisting of the closure of $C_0^\infty(S)$ with respect to the norm (3.13). Notice that $\mathring{W}_2^l(S)$ contains each function in $L^2(T)$ which, together with its derivatives of order $\leq l$, is square-integrable and equal to 0 outside the domain $S^{-\delta}$ for some $\delta > 0$.

Functions in this class are convolutions $v_n = w_n * v$ with an approximate identity sequence of non-negative symmetric functions $w_n \in C_0^\infty(\mathbb{R}^d)$,

$$w_n * v(t) = \int v(s)\tau_{-s}w_n(t)\,ds = \int w_n(s)\tau_{-s}v(t)\,ds, \qquad t \in T,$$

with derivatives $D^k(w_n * v) = D^k w_n * v$ in $L^2(T)$. The sequence $v_n \in \mathring{W}_2^l(S)$ belongs to the closure in $D(T)$ of the space $C_0^\infty(S)$ under the norm $\|u\|_L = \|Lu\|$ and approximates the original function $v \in D(T)$—these functions are bounded and converge weakly in the space $D(T)$ to v. In particular, as in the proof of §1, Lemma 2, we have, by (3.12),

$$\|v_n\|_L = \|L(w_n * v)\| \leq \int w_n(s)\|L\tau_{-s}v\|\,ds \leq C$$

and

$$(u, w_n * v) \to (u, v), \qquad u \in C_0^\infty(T).$$

As we have chosen it, the generalized function (3.2) is local, so that by the theorem in §1 its conjugate generalized function (3.8) is Markov. Under the assumption that $T = \mathbb{R}^d$ and (3.12) holds, we have established the duality property in connection with all domains $S \subseteq \mathbb{R}^d$, giving us the following proposition.

Lemma. Let the differential form (3.3) satisfy conditions (3.5) and (3.12). Then the generalized function (3.8) is Markov with respect to all domains $S \subseteq \mathbb{R}^d$.

Let us look at a stationary generalized function with a spectral density $f(\lambda)$ such that $1/f(\lambda)$ is a polynomial. It is biorthogonal to the stationary generalized function associated with the differential form (3.4), with differential operator $\mathscr{P} = \mathscr{P}(D)$ corresponding to the polynomial $\mathscr{P}(-i\lambda) = 1/f(\lambda)$, cf. (2.25). From (3.6) we have (3.5) satisfied when the spectral density $f(\lambda)$ is bounded:

$$f(\lambda) \leq 1/c.$$

Condition (1.18) will also hold because in the stationary case

$$\|L\tau_s v\| = \|Lv\|, \qquad s \in \mathbb{R}^d,$$

for all $v \in C_0^\infty(\mathbb{R}^d)$ and this equality extends to all functions $v \in D(\mathbb{R}^d)$ which are limits of functions in $C_0^\infty(\mathbb{R}^d)$ under the norm $\|v\|_L = \|Lv\|$.

We summarize this in the following result.

Theorem 1. *A stationary generalized function with bounded spectral density* $f(\lambda)$ *such that* $1/f(\lambda)$ *is a polynomial is Markov with respect to all open domains* $S \subseteq \mathbb{R}^d$.

It is worth pointing out here that when $f(\lambda)^{-1}$ is a function of the usual type, not connected with any polynomial, i.e., when

$$0 < c_1 \leq f(\lambda)^{-1} \leq c_2(1 + |\lambda|^2)^l, \qquad \lambda \in \mathbb{R}^d, \tag{3.13a}$$

then the duality condition for the related stationary biorthogonal functions is satisfied with respect to all domains $S \subseteq \mathbb{R}^d$. Indeed, condition (3.13a) guarantees the existence of the biorthogonal field Lu (see (2.19), the boundedness of the inverse operator L^{-1}, and the inequality

$$\|Lu\|^2 = \int |\tilde{u}|^2 f(\lambda)\, d\lambda \leq C \sum_{|k| \leq l} \|D^k u\|^2, \qquad u \in C_0^\infty(T).$$

As we have already seen, the following inequality is true for any differential form (3.3) of order $2l$:

$$\|Lu\|^2 = B^*(u, u) \leq C \sum_{|k| \leq l} \|D^k u\|^2, \qquad u \in C_0^\infty(T).$$

Suppose the reverse inequality holds for some constant $c > 0$:

$$\|Lu\|^2 = B^*(u, u) \geq c \sum_{|k| \leq l} \|D^k u\|^2, \qquad u \in C_0^\infty(T). \tag{3.14}$$

For example, notice that the inequality (3.14) is satisfied if the symmetric version of the differential form (3.3) is given by coefficients $b_{kj}(t)$, $|k|, |j| \leq l$, for which

$$\sum_{|k|, |j| \leq l} b_{kj}(t) x_k \overline{x_j} \geq c \sum_{|k| \leq l} |x_k|^2,$$

i.e., the positive definite quadratic form on the left-hand side is nondegenerate in the scalar variables x_k, $|k| \leq l$.

For a differential form (3.4) given by an operator $\mathscr{P} = \mathscr{P}(D)$ with constant coefficients, inequality (3.14) will hold if the corresponding polynomial $\mathscr{P}(-i\lambda)$ satisfies

$$\mathscr{P}(-i\lambda) \geq c(1 + |\lambda|^2)^l, \qquad \lambda \in \mathbb{R}^d,$$

for some $c > 0$ (cf. (3.6)).

Inequality (3.14) will be satisfied for a differential form (3.4) which satisfies (3.5) if the domain T is bounded and the operator $\mathscr{P}$ of order $2l$ has infinitely differentiable coefficients $b_k(t)$, $|k| \leq 2l$, satisfying the strong ellipticity condition in some neighborhood of the closure of the (open) domain $T \subseteq \mathbb{R}^d$:

$$(-1)^l \operatorname{Re} \sum_{|k| \leq l} b_k(t) \lambda^k > 0, \qquad \lambda \in \mathbb{R}^d.$$

In particular, under the strong ellipticity condition Gårding's inequality holds:

$$(\mathscr{P}u, u) + c_0(u, u) \geq c \sum_{|k| \leq l} \|D^k u\|^2, \qquad u \in C_0^\infty(T); \tag{3.15}$$

in view of nondegeneracy condition (3.5), this is equivalent to (3.14).

Under condition (3.14) we get equivalence of the norms

$$\|Lu\| = \|u\|_L \asymp \|u\|_l, \qquad u \in C_0^\infty(T);\dagger$$

thus the space $D(T)$—the closure of $C_0^\infty(T)$ under $\|u\|_L$—consists of the same functions as the Sobolev space $\mathring{W}_2^l(T)$—the closure of $C_0^\infty(T)$ under the norm (3.13). When $T = \mathbb{R}^d$ condition (3.12) will be satisfied because

$$\|\tau_s v\|_L^2 \asymp \sum_{|k| \leq l} \|D^k(\tau_s v)\|^2 = \sum_{|k| \leq l} \|D^k v\|^2 \asymp \|v\|_L^2$$

for variables $v \in D(T)$, $s \in \mathbb{R}^d$. We thus arrive at the following result, a companion to Theorem 1.

Theorem 2. *A generalized function* (3.8) *on a domain* $T = \mathbb{R}^d$ *which is associated with a differential form satisfying* (3.14) *is Markov with respect to all domains* $S \subseteq \mathbb{R}^d$.

Suppose that for every bounded domain $T_{\text{loc}} \subseteq T$ the inequality

$$\|Lu\|^2 = B^*(u, u) \geq c \sum_{|k| \leq l} \|D^k u\|_{\text{loc}}^2, \qquad u \in C_0^\infty(T), \tag{3.16}$$

holds, where the constant $c > 0$ depends on the domain T_{loc} and

$$\|u\|_{\text{loc}}^2 = \int_{T_{\text{loc}}} |u(t)|^2 \, dt, \qquad u \in L^2(T).$$

Under condition (3.16), a differential form (3.3) will have the property that for any function $w \in C_0^\infty(\mathbb{R}^d)$

$$\|L(w \cdot u)\|^2 = B^*(wu, wu) \leq cB^*(u, u) = c\|Lu\|^2, \qquad u \in C_0^\infty(T), \tag{3.17}$$

and we know this guarantees the duality of the biorthogonal fields (3.2) and (3.8) with respect to all domains $S \subseteq T$ which are bounded or have bounded complements in T. See §1, Lemma 1. In fact, for any function $w \in C_0^\infty(\mathbb{R}^d)$ and bounded domain T_{loc} containing the intersection Supp $w \cap T$, we have, under (3.16),

$$\|L(w \cdot u)\|^2 \leq C\|w \cdot u\|_l^2 \leq C_1 \sum_{|k| \leq l} \|D^k u\|_{\text{loc}}^2 \leq C_2\|Lu\|, \qquad u \in C_0^\infty(T).$$

As we shall see below, condition (3.16) is easy to check in several important cases.

Turning to the Markov condition, we formulate our results as follows.

† Recall that $\alpha \asymp \beta$ means that there exist constants c_1, c_2 such that $0 < c_1 \leq \alpha/\beta \leq c_2 < \infty$.

Theorem 3. *Given an arbitrary domain $T \subseteq \mathbb{R}^d$, a generalized function (3.8) associated with a differential form (3.3) which satisfies condition (3.16) or the weaker condition (3.17) is Markov with respect to all open domains $S \subseteq T$ which are bounded or have bounded complements in T.*

§4. Markov Random Functions Generated by Elliptic Differential Forms

1. Levy Brownian Motion

This term refers to the Gaussian random function $\xi(t)$, $t \in \mathbb{R}^d$, with mean zero and covariance

$$(\xi(s), \xi(t)) = \tfrac{1}{2}(|t| + |s| - |t - s|), \qquad s, t \in \mathbb{R}^d;$$

the following property is characteristic of it:

$$\|\xi(t) - \xi(s)\|^2 = |t - s|.$$

Using spherical coordinates, one can compute that

$$\int_{\mathbb{R}^d} \frac{|\exp(i\lambda t) - 1|^2}{|\lambda|^{2l}} \, d\lambda = (2\pi)^d \sigma^{-2} |t|,$$

where σ^2 is a constant and

$$l = \frac{d + 1}{2}.$$

Using this integral, one has the representation

$$\xi(t) = \frac{\sigma}{(2\pi)^{d/2}} \int \frac{\exp(i\lambda t) - 1}{|\lambda|^l} \, \eta(d\lambda), \qquad t \in \mathbb{R}^d,$$

where $\eta(d\lambda)$, $\lambda \in \mathbb{R}^d$, is the Gaussian orthogonal measure,

$$E|\eta(d\lambda)|^2 = d\lambda.$$

In fact, a function given this way has mean zero and

$$\|\xi(t) - \xi(s)\|^2 = \frac{\sigma^2}{(2\pi)^d} \int \frac{|\exp(i\lambda(t - s) - 1|^2}{|\lambda|^{2l}} \, d\lambda = |t - s|,$$

from which it follows that

$$\begin{aligned}
(\xi(t), \xi(s)) &= \tfrac{1}{2}(\|\xi(t)\|^2 + \|\xi(s)\|^2 - \|\xi(t) - \xi(s)\|^2) \\
&= \tfrac{1}{2}(|t| + |s| - |t - s|).
\end{aligned}$$

We will consider Levy Brownian motion as the generalized function

$$(u, \xi) = Ju = \int u(t)\xi(t)\, dt, \qquad u \in C_0^\infty(T),$$

in the domain

$$T = \mathbb{R}^d \setminus \{0\},$$

with a deleted point $t = 0$. We remark in connection with this that $\xi(t)$, $t \in \mathbb{R}^d$, is a real-valued function,

$$\xi(0) = 0,$$

and for any domain $S \subseteq \mathbb{R}^d$ the closed linear span $H(S)$ of the variables $\xi(t)$, $t \in S$, coincides with the closed linear span of the variables (u, ξ), Supp $u \subseteq S \setminus \{0\}$. With this the space $H(T) = H(\mathbb{R}^d)$ consists of all variables of the form

$$\eta = \int \varphi(\lambda)\eta(d\lambda), \qquad \varphi \in L^2(\mathbb{R}^d),$$

where $L^2(\mathbb{R}^d)$ is the usual space of measurable, square-integrable functions $\varphi(\lambda)$, $\lambda \in \mathbb{R}^d$, since the set of functions $[\exp(i\lambda t) - 1]/|\lambda|^l$ with parameter $t \in \mathbb{R}^d$ is complete in $L^2(\mathbb{R}^d)$. Indeed, if

$$\int \frac{\exp(i\lambda t) - 1}{|\lambda|^l} \overline{\varphi(\lambda)}\, d\lambda = 0, \qquad t \in \mathbb{R}^d,$$

then by taking the difference of these integrals for parameter values $t + s$ and t we get

$$\int \exp(i\lambda t) \left[\frac{\exp(i\lambda s) - 1}{|\lambda|^l} \overline{\varphi(\lambda)} \right] d\lambda = 0,$$

from which it follows that almost everywhere

$$\frac{\exp(i\lambda s) - 1}{|\lambda|^l} \overline{\varphi(\lambda)} = 0, \qquad \varphi(\lambda) = 0.$$

The generalized Levy–Brownian motion has the biorthogonal function

$$(u, \xi^*) = Lu = \frac{1}{\sigma(2\pi)^{d/2}} \int |\lambda|^l \tilde{u}(\lambda)\eta(d\lambda),$$

where

$$\tilde{u}(\lambda) = \int \exp(i\lambda t)u(t)\, dt, \qquad u \in C_0^\infty(T).$$

In fact, we have

$$(Ju, Lv) = \int u(t) \left[\overline{\frac{1}{(2\pi)^d} \int (\exp(-i\lambda t) - 1)\tilde{v}(\lambda)\, d\lambda} \right] dt$$

$$= \int u(t)\overline{v(t)}\, dt$$

for all $u, v \in C_0^\infty(T)$ since

$$\frac{1}{(2\pi)^d} \int \tilde{v}(\lambda)\, d\lambda = v(0) = 0.$$

Moreover,

$$H^*(T) = H(T),$$

since the collection of functions

$$|\lambda|^l \tilde{u}(\lambda), \qquad u \in C_0^\infty(T),$$

is complete in the space $L^2(\mathbb{R}^d)$: for $\varphi(\lambda) \in L^2(\mathbb{R}^d)$ the equation

$$\int \tilde{u}(\lambda)[\,|\lambda|^l\overline{\varphi(\lambda)}]\, d\lambda = 0, \qquad u \in C_0^\infty(T), \ T = \mathbb{R}^d \backslash \{0\},$$

means that the generalized function on the left side of the equation has as its support the point $t = 0$ and thus its Fourier transform is a polynomial $\mathscr{P}(\lambda)$ in $\lambda \in \mathbb{R}^d$ and $\overline{\varphi(\lambda)} = \mathscr{P}(\lambda)/|\lambda|^l$ almost everywhere. Since $2l = d + 1$ and $|\varphi(\lambda)|^2$ is integrable "at infinity," this polynomial must be a constant, $\mathscr{P}(\lambda) = c$, and by the integrability of $|\varphi(\lambda)|^2$ in a neighborhood of $\lambda = 0$, we must have $\mathscr{P}(0) = 0$, giving $\mathscr{P}(\lambda) = 0$, $\varphi(\lambda) = 0$ almost everywhere. Thus the biorthogonality conditions (1.4) and (1.5) are satisfied.

We will show that the duality condition (1.18) is also satisfied. We will make use of the representation

$$u(t) = \frac{1}{(2\pi)^d} \int \frac{\exp(-i\lambda t) - 1}{|\lambda|^l} \varphi(\lambda)\, d\lambda, \qquad t \in \mathbb{R}^d,$$

for $u \in C_0^\infty(T)$, where

$$\varphi(\lambda) = |\lambda|^l \tilde{u}(\lambda) \in L^2(\mathbb{R}^d), \qquad \|\varphi\| = \sigma(2\pi)^{d/2}\|Lu\|.$$

We have

$$|u(t)| \le C \left\| \frac{\exp(-i\lambda t) - 1}{|\lambda|^l} \right\| \|\varphi\|, \qquad t \in \mathbb{R}^d,$$

and obviously for any bounded domain $T_{\text{loc}} \subseteq T$

$$\|u\|_{\text{loc}} \le C_0 \|Lu\|.$$

By looking at the derivatives

$$D^k u(t) = \frac{1}{(2\pi)^d} \int \exp(-i\lambda t) \frac{(-i\lambda)^k}{|\lambda|^l} \varphi(\lambda)\, d\lambda, \qquad t \in \mathbb{R}^d,$$

and noticing that the function $(-i\lambda)^k/|\lambda|^l$, $0 < |k| \le l$, is square integrable in the domain $|\lambda| \le 1$ and bounded in the domain $|\lambda| > 1$, we arrive at the inequality

$$\|D^k u(t)\|_{\text{loc}} \le \left\| \frac{1}{(2\pi)^d} \int_{|\lambda| \le 1} \exp(-i\lambda t) \frac{(-i\lambda)^k}{|\lambda|^l} \varphi(\lambda)\, d\lambda \right\|_{\text{loc}}$$

$$+ \left\| \frac{1}{(2\pi)^d} \int_{|\lambda| > 1} \exp(-i\lambda t) \frac{(-i\lambda)^k}{|\lambda|^l} \varphi(\lambda)\, d\lambda \right\|_{\text{loc}}$$

$$\le C_k \|Lu\|, \qquad 0 < |k| \le l.$$

In other words, we get (3.16),

$$\|Lu\|^2 \ge c \sum_{|k| \le l} \|D^k u\|_{\text{loc}}^2, \qquad u \in C_0^\infty(T),$$

and we know this makes the duality condition (1.18) true.

It is clear that the operator L is local in the case where the dimension d is odd, making $l = (d + 1)/2$ an integer, because

$$(Lu, Lv) = \frac{1}{\sigma^2 (2\pi)^d} \int |\lambda|^{2l} \tilde{u}(\lambda) \overline{\tilde{v}(\lambda)}\, d\lambda = \frac{1}{\sigma^2} (-\Delta^l u, v), \qquad u, v \in C_0^\infty(T),$$

where $\Delta = \sum_k \partial^2/\partial t_k^2$ is the Laplacian.

Moreover, this local property does not hold when the dimension d is even and l is not an integer.

In fact, if a bilinear form is of the type

$$\int \tilde{u}(\lambda) \overline{\tilde{v}(\lambda)} \varphi(\lambda)\, d\lambda, \qquad u, v \in C_0^\infty(T),$$

with $\varphi(\lambda)$, $\lambda \in \mathbb{R}^d$, locally-integrable and of power-law growth, and if the form vanishes for all u, v with disjoint supports then this implies, as we saw earlier when studying stationary functions, that $\varphi(\lambda)$ is a generalized Fourier transform of a generalized function whose support is the single point $t = 0$, and thus $\varphi(\lambda)$ must be a polynomial in $\lambda \in \mathbb{R}^d$. In the case of Levy Brownian motion on an even-dimensional space $\mathbb{R}^d$, we are dealing with the bilinear form

$$(Lu, Lv) = \frac{1}{\sigma^4} \int \tilde{u}(\lambda) \overline{\tilde{v}(\lambda)} |\lambda|^{d+1}\, d\lambda;$$

but for $d + 1$ odd, $|\lambda|^{d+1}$ is not a polynomial.

Applying the general theorem in §1, we arrive at the following conclusion:

If the dimension d is odd, the Levy Brownian motion on the space $\mathbb{R}^d$ is Markov with respect to all domains $S \subseteq T$ which are bounded or have bounded complements in T (of course, here one can substitute the domain $T = \mathbb{R}^d \backslash \{0\}$ for the whole space $\mathbb{R}^d$). When the dimension of $\mathbb{R}^d$ is even, then it does not have the Markov property.

With subsequent results in mind we point out that the differential form

$$(Lu, Lv) = \frac{1}{\sigma^2}(-\Delta^l u, v), \qquad u, v \in C_0^\infty(T),$$

is elliptic.

2. Structure of Spaces for Given Elliptic Forms

Let the generalized random function

$$(u, \xi) = Ju, \qquad u \in C_0^\infty(T), \tag{4.1}$$

have biorthogonal function

$$(u, \xi^*) = Lu, \qquad u \in C_0^\infty(T), \tag{4.2}$$

with a covariance functional of the form

$$(Lu, Lv) = B^*(u, v) = (\mathscr{P}u, v),$$

where

$$\mathscr{P}u(t) = \sum_{|k| \leq 2l} b_k(t) D^k u(t), \qquad t \in T, \tag{4.3}$$

is a linear differential operator with sufficiently smooth, bounded coefficients. The bilinear differential form $B^*(u, v)$, $u, v \in C_0^\infty(T)$, is called *elliptic* if it satisfies condition (3.16); in this case the norm $\|u\|_L = \|Lu\|$ satisfies the inequality

$$c \sum_{|k| \leq l} \|D^k u\|_{\text{loc}}^2 \leq \|u\|_L^2 \leq C \sum_{|k| \leq l} \|D^k u\|^2, \qquad u \in C_0^\infty(T), \tag{4.4}$$

where the constant $c > 0$ on the left generally depends on the associated bounded domain $T_{\text{loc}} \subseteq T$, to which we also associate the norm

$$\|u\|_{\text{loc}}^2 = \int_{T_{\text{loc}}} |u(t)|^2 \, dt, \qquad u \in L^2(T).$$

As we have seen, under condition (4.4), the biorthogonal generalized functions (4.1), (4.2) are dual with respect to all domains $S \subseteq T$ which are bounded or have bounded complement in T, and the random function (4.1) is Markov with respect to these domains. See, in this connection, §3, Theorem 3.

We consider the space $V(T)$ gotten by completing $C_0^\infty(T)$ with respect to the norm $\|u\|_L$. For an arbitrary element $v \in V(T)$, we know

$$v = (u, v), \qquad u \in C_0^\infty(T),$$

is a generalized function which is the limit of some sequence $v_n \in C_0^\infty(T)$. By (4.4), for any bounded domain T_{loc} we have

$$\|u\|_L \asymp \|u\|_l, \qquad u \in C_0^\infty(T_{\text{loc}}); \tag{4.5}$$

the norm on the right is associated with the space $\mathring{W}_2^l(T)$,

$$\|u\|_l^2 = \sum_{|k| \leq l} \|D^k u\|^2, \qquad u \in \mathring{W}_2^l(T). \tag{4.6}$$

We also recall that for any function $w \in C_0^\infty(\mathbb{R}^d)$ we have inequality (3.17):

$$\|w \cdot u\|_L \leq C\|u\|_L, \qquad u \in C_0^\infty(T).$$

We take a real function $w \in C_0^\infty(\mathbb{R}^d)$ with $w(t) = 1$ for $t \in T_{\text{loc}}$. It is clear that the sequence wv_n, $n = 1, 2, \ldots$, is Cauchy in the space $V(T)$,

$$\|w(v_n - v_m)\|_L \leq C\|v_n - v_m\|_L \to 0, \qquad n, m \to \infty,$$

and

$$(u, w \cdot v_n) = (u, v_n) \to (u, v), \qquad u \in C_0^\infty(T_{\text{loc}}).$$

On the other hand, by (4.5) the sequence wv_n, $n = 1, 2, \ldots$, converges in the space $\mathring{W}_2^l(T)$ to a function $\tilde{v}(t)$, $t \in T$, for which

$$(u, \tilde{v}) = (u, v), \qquad u \in C_0^\infty(T_{\text{loc}}).$$

The resulting equation shows that a *generalized function* $v \in V(T)$ *can be represented on any bounded domain* $T_{\text{loc}} \subseteq T$ *as a generalized function in the class* $\mathring{W}_2^l(T)$.

Later we shall need some well-known properties of the space $\mathring{W}_2^l(T)$—the closure of the space $C_0^\infty(T)$ under the norm (4.6).

Every function $v \in \mathring{W}_2^l(T)$ has generalized derivatives

$$D^k v \in L^2(T), \qquad |k| \leq l. \tag{4.7}$$

To see this, if $v = \lim u_n$ is the limit of $u_n \in C_0^\infty(T)$ under the norm (4.6) then the limits

$$D^k v = \lim_{n \to \infty} D^k u_n, \qquad |k| \leq l,$$

exist in $L^2(T)$ and

$$\int u(t) D^k \overline{v(t)}\, dt = \lim \int u(t) D^k \overline{u_n(t)}\, dt = \lim (-1)^{|k|} \int D^k u(t) \overline{u_n(t)}\, dt$$

$$= (-1)^{|k|} \int D^k u(t) \overline{v(t)}\, dt, \qquad u \in C_0^\infty(T).$$

Clearly, the space $\mathring{W}_2^l(T)$ is invariant under multiplication by functions $w \in C_0^\infty(\mathbb{R}^d)$: if $v = \lim u_n$ with $u_n \in C_0^\infty(T)$, then $vw = \lim u_n w$ under the norm (4.6).

Consider an arbitrary function $v \in \mathring{W}_2^l(T)$. It can be approximated in the space $\mathring{W}_2^l(T)$ by functions v_n with compact support

$$\text{Supp } v_n \subseteq \text{Supp } v.$$

For example, one can take $v_n = v \cdot w(t/n)$, where $w \in C_0^\infty(\mathbb{R}^d)$ and $w(t) = 1$ for $|t| \leq 1$; for such a sequence

$$\|v - v_n\|_l^2 \leq C \sum_{|k| \leq l} \int_{|t| > n} |D^k v(t)|^2 \, dt \to 0$$

as $n \to \infty$.

To complete this, notice that if the support of v lies in the open domain $S \subseteq T$, $\text{Supp } v \subseteq S$, then $v \in \mathring{W}_2^l(S)$. In fact, without loss of generality we can assume that the function $v \in \mathring{W}_2^l(T)$ has compact support; but then it can be approximated by convolution $v * w_n \in C_0^\infty(S)$ with a standard approximate identity sequence $w_n \in C_0^\infty(\mathbb{R}^d)$ because for all the derivatives $D^k v$, $|k| \leq l$, we have

$$D^k(v * w_n) = D^k v * w_n \to D^k v \quad \text{in } L^2(T)$$

as $n \to \infty$.

It is evident from the discussion above that $\mathring{W}_2^l(\mathbb{R}^d)$ is the collection of all functions $v(t)$, $t \in \mathbb{R}^d$, which have generalized derivates $D^k v \in L^2(\mathbb{R}^d)$, $|k| \leq l$.

We call an open domain $S \subseteq \mathbb{R}^d$ locally star-like if each point s on the boundary $\Gamma = \partial S$ has a neighborhood s^ε such that the similarity transformation with respect to some point t_0 in the domain $S \cap s^\varepsilon$,

$$t \rightsquigarrow t_0 + r(t - t_0), \qquad 0 < r < 1,$$

takes the closure $\overline{S \cap s^\varepsilon}$ into the interior of this domain.

We will look at a locally star-like domain $S \subseteq \mathbb{R}^d$ and a function $v \in \mathring{W}_2^l(S)$. Because it is the limit of some sequence $u_n \in C_0^\infty(S)$, the function $v(t)$, $t \in \mathbb{R}^d$, has the property that

$$v(t) \in \mathring{W}_2^l(\mathbb{R}^d), \qquad v(t) = 0 \quad \text{for } t \in \mathbb{R}^d \setminus \bar{S}. \tag{4.8}$$

We shall show that this condition defines a function $v \in \mathring{W}_2^l(S)$.

We suppose that the boundary Γ of the domain S in $\mathbb{R}^d$ is finite. We take an open covering by the associated neighborhoods s^ε and a "partition of unity"

$$1 = \sum_{k=1}^n u_k(t), \qquad t \in \Gamma^\delta, \tag{4.9}$$

where each function $u_k \in C_0^\infty(\mathbb{R}^d)$ is zero outside some neighborhood s^ε and Γ^δ is a sufficiently small neighborhood of the boundary Γ. We have

$$v = v\left(1 - \sum_{k=1}^n u_k\right) + \sum_{k=1}^n v \cdot u_k,$$

with each term in the sum a function in $\mathring{W}_2^1(\mathbb{R}^d)$. Moreover the first term, $v_0 = v(1 - \sum_{k=1}^n u_k)$, has support Supp $v_0 \subseteq S^{-\delta}$ and thus belongs to the space $\mathring{W}_2^1(S)$; the remaining terms $v_k = v \cdot u_k$ vanish outside corresponding neighborhoods s^ε and satisfy (4.8). Under a suitable similarity transformation with coefficient $r < 1$, the function v_k is transformed into a function $v_{kr} \in \mathring{W}_2^1(\mathbb{R}^d)$ with compact support strictly contained in the interior of the domain S; thus $v_{kr} \in \mathring{W}_2^1(S)$. It is clear that as $r \to 1$, $v_{kr} \to v_k$ in the norm (4.6) and $v_k \in \mathring{W}_2^1(S)$. In other words,

$$v = v_0 + \sum_{k=0}^n v_k \in \mathring{W}_2^1(S),$$

which is what we wished to prove. In the case of an infinite boundary, one can replace the original function $v \in \mathring{W}_2^1(\mathbb{R}^d)$ from the very beginning by functions approximating v, of the form $v \cdot w$, $w \in C_0^\infty(\mathbb{R}^d)$, with compact supports which have boundaries in common with S only on finite portions of $\Gamma = \partial S$.

Recall that if a locally integrable function $f(t)$, $t \in T$, has a generalized derivative $g(t)$ of this type, say,

$$-\int \frac{\partial}{\partial t_i} u(t) \overline{f(t)} \, dt = \int u(t) \overline{g(t)} \, dt, \qquad u \in C_0^\infty(T),$$

then for almost all values of the variables t_j, $j \neq i$, the function $f(t)$ is absolutely continuous in the variable t_i and has the usual derivative $(\partial/\partial t_i) f(t) = g(t)$ for almost all t_i. (More precisely, there exists an equivalent function having the indicated properties.)† For instance, from the equation derived above, it immediately follows that in all sufficiently small neighborhoods $|t_k - t_k^0| < \varepsilon$, $k = 1, \ldots, d$, in the domain $T \subseteq \mathbb{R}^d$ and for almost all fixed t_j, $j \neq i$, the difference

$$f(t) - \int_{t_i^0}^{t_i} g(t) \, dt_i = C$$

is constant for almost all t_i. Conversely, if a function $f(t)$ is absolutely continuous in t_i in the sense above, then $g(t) = (\partial/\partial t_i) f(t)$ will be its generalized derivative. Moreover, if the same absolute continuity in the variable t_j holds for both $f(t)$ and $g(t) = (\partial/\partial t_i) f(t)$, then $(\partial/\partial t_j) g(t)$ will be the generalized derivative of the function $(\partial/\partial t_i) f(t)$ with respect to the variable t_i because

$$\int u(t) \frac{\partial}{\partial t_j} g(t) \, dt = -\int \frac{\partial}{\partial t_j} u(t) \frac{\partial}{\partial t_i} \overline{f(t)} \, dt$$

$$= \int \frac{\partial^2}{\partial t_i \, \partial t_j} u(t) f(t) \, dt = -\int \frac{\partial}{\partial t_i} u(t) \frac{\partial}{\partial t_j} f(t) \, dt, \, u \in C_0^\infty(T).$$

† Similar properties which we will consider below also carry over to a suitable representative from the class of equivalent functions, i.e., functions which coincide almost everywhere.

As a consequence we get that for almost all t_k, $k \neq i$, the function $(\partial/\partial t_j)f(t)$ is absolutely continuous in the variable t_i and has the usual derivative

$$\frac{\partial}{\partial t_i}\left[\frac{\partial}{\partial t_j}f(t)\right] = \frac{\partial}{\partial t_j}g(t). \tag{4.10}$$

Recall here also that the generalized derivative $D^k f(t)$,

$$\int u(t)D^k f(t)\,dt = (-1)^{|k|}\int D^k u(t)f(t)\,dt, \qquad u \in C_0^\infty(T),$$

does not depend on the order of differentiation since the derivative $D^k u(t)$ does not.

Further notice that the above property of absolute continuity of the function f and its generalized derivatives $D^k f$, $|k| \leq l$, is preserved under nondegenerate l-times continuously differentiable changes of variable. Roughly speaking, under such a transformation of the variables, with Jacobean $\mathscr{J}$, the function f, understood in the generalized sense, is carried into the function $f \cdot \mathscr{J}$.

We already noted that a function $v \in \mathring{W}_2^l(T)$ has generalized derivatives $D^k v \in \mathscr{L}^2(T)$, $|k| \leq l$. The presence of these generalized derivatives means that any of the functions $D^k v(t)$, $|k| \leq l - 1$, arising from the function v, is absolutely continuous in each coordinate t_i of the variable $t = (t_1, \ldots, t_d)$ for almost all values of $t_j, j \neq i$; furthermore, the derivative $(\partial/\partial t_i)D^k v(t)$ coincides with the corresponding generalized derivative. This property is preserved under nondegenerate, l-times continuously differentiable transformations of the coordinates, which for an l-smooth boundary Γ of the domain $S \subseteq \mathbb{R}^d$ allows us to express condition (4.8) in the form of the following boundary conditions:

$$\frac{\partial^k}{\partial r^k}v(s) = 0, \qquad k \leq l - 1, \tag{4.11}$$

for almost all $s \in \Gamma$. Here it is assumed that in a small neighborhood of each point $s_0 \in \Gamma$, there exists a local coordinate system

$$t = (s, r), \qquad s \in \Gamma_{\mathrm{loc}}, \qquad -\varepsilon < r < \varepsilon, \tag{4.12}$$

in which r parametrizes motion along any curve not tangent to Γ and passing through the point $s_0 = (s_0, 0)$—for instance, r could parametrize along an exterior normal to Γ at the point s_0. As noted above, for almost all $s \in \Gamma$ there exist absolutely continuous derivatives $(\partial^k/\partial r^k)v(t)$, $k \leq l - 1$, with values at $t = s$ indicated in the boundary conditions (4.11).

Clearly these conditions are filled for any function $v \in \mathring{W}_2^l(S)$, since $v(s, r) = 0$ for $r > 0$ and $v(s, r)$, together with all of its derivatives $(\partial^k/\partial r^k)v(s, r)$, $k \leq l - 1$, is continuous in r.

We will show that if a function $v \in \mathring{W}^l_2(\mathbb{R}^d)$ satisfies boundary conditions (4.11) then its "patch"

$$v(t) = \begin{cases} v(t), & t \in \bar{S}, \\ 0, & t \in \mathbb{R}^d \setminus \bar{S}, \end{cases} \tag{4.13}$$

will be a function in the class $\mathring{W}^l_2(S)$.

For the proof we will consider first a Γ_{loc} which has a "flat piece" where, let us say, the coordinates (4.12) have the form

$$s = (t_1, \ldots, t_{d-1}, 0), \qquad r = t_d.$$

We will show that the function $v(t)$ defined by formula (4.13), together with its derivatives $D^k v(t)$, $|k| \le l - 1$, is absolutely continuous in each variable t_i, for almost all fixed t_j, $j \ne i$, in a neighborhood (4.12) of the given type. With regard to the variables $t_1, \ldots, t_{d-1}$ this is obvious because $D^k \tilde{v}(t) = D^k v(t)$ or $D^k \tilde{v}(t) = 0$ for all $r \ne 0$. For the variable $t_d = r$, it is also clear for the unmixed partials $(\partial^k / \partial r^k) \tilde{v}(t)$, $k \le l - 1$, by the nature of the function v and the patch $\tilde{v}$. One can convince oneself of the absolute continuity of the remaining partials by looking sequentially at higher derivatives in the variable t_j, $j \ne d$, of the functions $(\partial^k / \partial r^k) \tilde{v}(t)$. We will demonstrate just the first step. The square-integrable functions

$$f(t) = \frac{\partial^{k-1}}{\partial r^{k-1}} \tilde{v}(t), \qquad g(t) = \frac{\partial^k}{\partial r^k} \tilde{v}(t), \qquad k \le l - 1,$$

are absolutely continuous in t_j and thus the derivative $(\partial / \partial t_j) f(t)$ is absolutely continuous in the variable $r = t_d$ for almost all t_k, $k \ne d$, and

$$\frac{\partial}{\partial r} \left[\frac{\partial}{\partial t_j} f(t) \right] = \frac{\partial}{\partial r} \left[\frac{\partial}{\partial t_j} \frac{\partial^{k-1}}{\partial r^{k-1}} \tilde{v}(t) \right] = \frac{\partial}{\partial t_j} g(t) = \frac{\partial}{\partial t_j} \frac{\partial^k}{\partial r^k} \tilde{v}(t), \qquad k \le l - 1,$$

cf. (4.10). The next step is to look at functions

$$f(t) = \frac{\partial}{\partial t_j} \frac{\partial^{k-2}}{\partial r^{k-2}} \tilde{v}(t), \qquad g(t) = \frac{\partial}{\partial t_j} \frac{\partial^{k-1}}{\partial r^{k-1}} \tilde{v}(t), \qquad k \le l - 1,$$

and so on, until one has the result for all derivatives $D^k v(t)$, $|k| \le l - 1$. The absolutely continuity of the derivatives $D^k v(t)$ established in the neighborhood (4.12) does not change under nondegenerate, l-times continuously differentiable changes of coordinates and this allows us to go from an arbitrary Γ_{loc} to the type of "flat piece" just considered. We already saw that we can assume that the boundary Γ of the domain $S \subseteq \mathbb{R}^d$ is finite. By taking a suitable partition of unity (4.9) we have the representation

$$\tilde{v} = \tilde{v} \cdot \left(1 - \sum_{k=1}^n u_k \right) + \sum_{k=1}^n \tilde{v} \cdot u_k,$$

in which each term has square-integrable generalized derivatives up to order l and is a function in the class $\mathring{W}^l_2(S)$. In short, we have $\tilde{v} \in \mathring{W}^l_2(S)$.

Thus the *existence of square integrable generalized derivatives* (4.7) *and zero boundary values* (4.11) *characterizes the class* $\overset{\circ}{W}^l_2(S)$ *for a domain* $S \subseteq \mathbb{R}^d$ *with l-smooth boundary* Γ.

This description of the class $\overset{\circ}{W}^l_2(S)$ extends to domains S with *piecewise l-smooth boundaries* Γ; these are domains which can be represented as the intersection $S = \bigcap S_k$ of a finite number of domains $S_k \subseteq \mathbb{R}^d$ with *l*-smooth boundaries Γ_k and having the property that in a sufficiently small neighborhood of a boundary point $s \in \Gamma_k$, local coordinate systems of the type (4.12) can be obtained from one another by nondegenerate *l*-smooth transformations.

According to the boundary conditions described above, we can regard a function $v \in \overset{\circ}{W}^l_2(T)$ as being equal to zero on the boundary, for a piecewise *l*-smooth domain $T \subseteq \mathbb{R}^d$. This allows us to make the following obvious modification of (4.8) for locally star-like domains $S \subseteq T$:

$$v \in \overset{\circ}{W}^l_2(T), \qquad v(t) = 0 \quad \text{for } T \backslash \bar{S}, \tag{4.14}$$

where $\bar{S}$ is the closure of S in T.

We apply this criterion for a function $v \in \overset{\circ}{W}^l_2(T)$ to belong to the class $\overset{\circ}{W}^l_2(S)$ to the case where the domains in question are $S_1 \subseteq T$ and $S_2 = T \backslash \bar{S}_1$, having a common boundary Γ in the domain T. We take a function $v \in \overset{\circ}{W}^l_2(T)$ satisfying boundary conditions (4.11) on $\Gamma \subseteq T$ and its patches of the type (4.13):

$$\tilde{v}_1(t) = \begin{cases} v(t), & t \in \bar{S}_1, \\ 0, & t \in T \backslash \bar{S}_1, \end{cases} \in \overset{\circ}{W}^l_2(S_1),$$

$$\tilde{v}_2(T) = \begin{cases} 0, & t \in T \backslash \bar{S}_2, \\ v(t), & t \in \bar{S}_2, \end{cases} \in \overset{\circ}{W}^l_2(S_2).$$

We get that

$$v(t) = \tilde{v}_1(t) + \tilde{v}_2(t) \in \overset{\circ}{W}^l_2(S_1 \cup S_2).$$

3. Boundary Conditions

Let us return to the spaces $V(T)$, $S \subseteq T$, introduced earlier in connection with the duality condition (1.18). As we saw, if inequality (4.4) holds, this condition is satisfied with respect to open domains $S \subseteq T$ which are bounded or have bounded complements in T; however, when one can take the whole domain T in place of T_{loc} in the equivalence relation (4.5), then condition (1.18) is valid for all open domains $S \subseteq T$. The space $V(S)$ was defined for arbitrary domains $S \subseteq T$ as the closure of all generalized functions

$$v \in V(T), \qquad \text{Supp } v \subseteq S^{-\varepsilon}, \qquad \varepsilon > 0,$$

where we recall that $S^{-\varepsilon} = T - \overline{(S^c)^\varepsilon}$.

Under the duality conditions for the domains $S \subseteq T$, the space $V(S)$ coincides with the closure of $C_0^\infty(S)$ with respect to a norm $\|u\|_L$ satisfying the equivalence relation (4.5) in every bounded domain $T_{\text{loc}} \subseteq T$. We already saw that in this case each function $v \in V(S)$ belongs locally to the class $\mathring{W}_2^l(S)$, and under condition (4.5) with T replacing T_{loc}, we see that the spaces $V(S)$ and $\mathring{W}_2^l(S)$ consist of exactly the same functions.

Let a domain $T \subseteq \mathbb{R}^d$ have piecewise l-smooth boundary. In the same way as in (4.14), for a locally star-like domain $S \subseteq T$, functions $v \in V(S)$ can be described in the following fashion:

$$v \in V(T), \qquad v(t) = 0 \quad \text{for } t \in T \backslash \bar{S}. \tag{4.15}$$

It is evident that to show this one needs only the case $v \in \mathring{W}_2^l(T)$ locally; in this situation condition (4.15) is true for domains $S \subseteq T$ with finite boundary $\Gamma = \partial S$ in T. Under (4.15), taking a function $w \in C_0^\infty(\mathbb{R}^d)$ with $w(t) = 1$ for $t \in \Gamma^\varepsilon$, we get

$$v \cdot w \in \mathring{W}_2^l(S) \subseteq V(S),$$

and since the difference $v - v \cdot w \in V(T)$ has support in the domain $S^{-\varepsilon}$, $v - v \cdot w \in V(S)$ and

$$v = v \cdot w + (v - v \cdot w) \in V(S).$$

As a consequence of condition (4.15) characterizing functions $v \in V(S)$ we have the equation

$$\bigcap_{\varepsilon > 0} V(S^\varepsilon) = V(S);$$

the intersection on the left clearly consists of functions $v \in V(T)$ which are zero outside the closure $\bar{S} = \bigcap_{\varepsilon > 0} S^\varepsilon$.

Recall that for domains S satisfying the duality condition (1.18), the space $V(S)$ is unitarily isomorphic to the associated space $H^*(S)$—we are referring to the correspondence (1.9). Applying this to the domain S^ε we get

$$H_+^*(S) = \bigcap_{\varepsilon > 0} H^*(S^\varepsilon) = H^*(S). \tag{4.16}$$

The analogous equation

$$H_+(S) = H(S) \tag{4.17}$$

is also true for the spaces $H(S)$, $S \subseteq T$, with this reservation: we are taking domains $S = S_1$ having a common boundary with the (locally star-like) complementary domains $S_2 = T \backslash \bar{S}_1$. In fact, from the duality condition (1.18) we get

$$H(S_1) = H_+^*(S_1^c)^\perp = H^*(S_2)^\perp = H_+(S_2^c) = H_+(S_1).$$

One can characterize property (4.17) as the *continuous renewal* of the random field $H(S)$, $S \subseteq T$, by the family of expanding domains S^ε, $\varepsilon > 0$.

Notice that with (4.17) true not only for $S_1 = S$ but also for the complementary domain $S_2 = T \backslash \bar{S}$ with boundary $\Gamma = \partial S$, we have

$$H_+(\Gamma_-) = H_+(\Gamma)$$

—see Chapter 2, (3.20)—and $H_+(\Gamma)$ is the minimal space splitting $H(S_1) = H_+(S_1)$ and $H(S_2) = H_+(S_2)$.

We formulate a summarizing result for a generalized function (4.1) and biorthogonal function (4.2) with elliptic bilinear form.

Theorem 1. *A Markov random field $H(S)$, $S \subseteq T$, generated by a generalized function (4.1), is continuously renewed by every family S^ε, $\varepsilon > 0$, in some neighborhood of a locally star-like domain $S \subseteq T$ which has a common boundary Γ in T with complementary domain $T \backslash \bar{S}$ of the same type; moreover, the boundary space*

$$H_+(\Gamma) = \bigcap_{\varepsilon > 0} H(\Gamma^\varepsilon)$$

is the minimal space which splits

$$H(S) = H_+(S), \qquad H(T \backslash \bar{S}) = H_+(T \backslash \bar{S}).$$

We can sharpen this by noting that under the usual ellipticity condition arising from (4.4), this theorem deals with domains $S \subseteq T$ which are bounded or have bounded complements in T; under even stronger conditions, where we have the equivalence condition (4.5) with T replacing T_{loc}, Theorem 1 is true for all domains $S \subseteq T$ which, together with their complementary domains $T \supseteq \bar{S}$, are locally star-like.

For one important application of Theorem 1, consider the following example.

EXAMPLE. Let $\xi(t)$, $t \in \mathbb{R}^1$ be a stationary random function with spectral density

$$f(\lambda) \asymp (1 + |\lambda|^2)^{-l}, \qquad \lambda \in \mathbb{R},$$

where l is a positive integer. The norm on the associated space $H(\mathbb{R})$ is equivalent to the norm which we had in the case of Markov stationary function with spectral density

$$f(\lambda) = (1 + |\lambda|^2)^{-l}, \qquad \lambda \in \mathbb{R},$$

and in examining the structure of the spaces $H(S)$, $S \subseteq \mathbb{R}$, we can take, without loss of generality, this particular case where, as we have seen, condition (4.5) is true with T_{loc} replaced by $T = \mathbb{R}$. For the Markov function we have taken there exist derivatives $\xi^{(k)}(t)$, $k \leq l - 1$, and it is well known that their closed linear span, including $\xi(t)$, splits the "past" $H(-\infty, t)$ and the "future" $H(t, \infty)$. Clearly their linear span is

$$\bigvee_{k=0}^{l-1} \xi^{(k)}(t) = H_+(t),$$

because the space

$$H_+(t) = \bigcap_{\varepsilon > 0} H(-\varepsilon, \varepsilon)$$

contains the derivatives $\xi^{(k)}(t)$ and is the minimal space which splits $H(-\infty, t)$ $= H_+(-\infty, t)$ and $H(t, \infty) = H_+(t, \infty)$.

Now we turn to the general case.

Let $S \subseteq T$ be a domain with piecewise l-smooth boundary $\Gamma = \partial S$ in T. We take a point $s_0 \in \Gamma$ for which there is a sufficiently small neighborhood on which we can introduce local coordinates (4.12) gotten from the original coordinates by an l-smooth transformation with nondegenerate Jacobean $\mathscr{J}$. We choose the neighborhood so that it contains the closure of a neighborhood of the form

$$\Gamma_{\text{loc}}^\varepsilon = \{t = (s, r), s \in \Gamma_{\text{loc}}, -\varepsilon < r < \varepsilon\},$$

where Γ_{loc} is some small piece of the boundary Γ. Later on it will essentially turn out that multiplication by the l-times continuously differentiable function $\mathscr{J}$ is bounded on the space $\mathring{W}_2^l(\Gamma_{\text{loc}}^\delta)$, so for simplicity of notation we will assume $\mathscr{J} = 1$. Our object is to define the "trace" of the generalized derivatives $(\partial^k/\partial r^k)\xi$, $k \leq l - 1$, of the generalized random function (4.1) on the boundary Γ.

Under condition (4.4) we have

$$\|Ju\| = \sup_{\|Lv\| = 1} |(Ju, Lv)| = \sup_{\|Lv\| = 1} |(u, v)| \leq \sup_{\|v\| \leq 1/c} |(u, v)|$$

$$\leq \frac{1}{c} \|u\|,$$

where the constant c depends on the bounded domain $T_{\text{loc}} \supseteq \text{Supp } u$. It is evident that the operator J is bounded in $L^2(T_{\text{loc}})$ norm and can be extended by continuity from $C_0^\infty(T_{\text{loc}}) \subseteq L^2(T_{\text{loc}})$ to the whole space $L^2(T_{\text{loc}})$. From this extension we have, for any variable $\eta \in H(T)$ and the function $v = v(\eta) \in V(T)$ associated with it by formula (1.8), the equation

$$(Ju, \eta) = \int u(t)\overline{v(t)}\, dt, \qquad u \in L^2(T_{\text{loc}}); \tag{4.18}$$

this arises from the relation

$$(Ju, Lv) = (u, v), \qquad u, v \in C_0^\infty(T),$$

by a limiting procedure for functions $u \in L^2(T_{\text{loc}})$ and $v \in V(T)$. Recall that the function $v(t)$ coincides on the bounded domain $T_{\text{loc}} \subseteq T$ with a function from the class $\mathring{W}_2^l(T)$.

We take an arbitrary function $u_{\text{loc}} \in C_0^\infty(\Gamma_{\text{loc}}^\varepsilon)$ with support in the domain $\Gamma_{\text{loc}}^\varepsilon = \Gamma_{\text{loc}} \times (-\varepsilon, \varepsilon)$ described above. The function $u_{\text{loc}}(s, r)$ is l-smooth in the variables s, r and vanishes on $|r| \geq \delta$ for some $\delta < \varepsilon$; we define the function u_{loc} for all r by setting it equal to zero for $|r| > \delta$. We let

$$u_n(s, r) = u_{\text{loc}}(s, nr) \cdot n, \qquad n = 1, 2, \ldots,$$

and consider the variables

$$(u_n, \zeta^{(k)}) = (-1)^k J u_n^{(k)},$$

where

$$u_n^{(k)}(s, r) = \frac{\partial^k}{\partial r^k} u_n(s, r), \qquad k \leq l - 1.$$

In view of (4.18) we have, for every variable $\eta \in H(T)$ and associated $v \in V(T)$,

$$(-1)^k (J u_n^{(k)}, \eta) = \int_{\Gamma_{\text{loc}}} \left[(-1)^k \int_{-\varepsilon}^\varepsilon u_n^{(k)}(s, r)\overline{v(s, r)}\, dr \right] ds$$

$$= \int_{\Gamma_{\text{loc}}} \int_{-\varepsilon}^\varepsilon u_n(s, r)\overline{v^{(k)}(s, r)}\, dr\, ds;$$

recall here that we assumed that the Jacobean of the transformation was equal to 1 and that for almost all s the function $v(s, r)$, which belongs locally to the class $\mathring{W}_2^l(T)$, has $l - 1$ absolutely continuous derivatives in r. Let

$$u(s) = \int u_{\text{loc}}(s, r)\, dr, \qquad s \in \Gamma_{\text{loc}}. \tag{4.19}$$

For any $\delta > 0$, for almost all $s \in \Gamma_{\text{loc}}$ and n large enough we have

$$\left| \int u_n(s, r)\overline{v^{(k)}(s, r)}\, dr - u(s)\overline{v^{(k)}(s, 0)} \right|$$

$$\leq C \sup_{|r| \leq \delta} |v^{(k)}(s, r) - v^{(k)}(s, 0)| \leq C \int_{-\delta}^\delta |v^{(k+1)}(s, r)|\, dr.$$

Clearly,

$$\int_{\Gamma_{\text{loc}}} \int_{-\delta}^\delta |v^{(k+1)}(s, r)|\, dr\, ds \to 0,$$

as $\delta \to 0$ because the function $v^{(k+1)}$, $k \leq l - 1$, is square-integrable and, once again, $v(t)$ coincides with a function in the class $\mathring{W}_2^l(T)$ for $t \in \Gamma_{\text{loc}}^\varepsilon$. We see that

$$\lim_n (u_n, \zeta^{(k)}) = \int_{\Gamma_{\text{loc}}} u(s)\overline{v^{(k)}(s, 0)}\, ds, \qquad \eta \in H(T),$$

where $v^{(k)}(s, 0) = v^{(k)}(s)$, $s \in \Gamma_{loc}$. Moreover, for $k \le l - 1$

$$|(u_n, \xi^{(k)})|^2 \le C_1 \int_{\Gamma_{loc}} |v^{(k)}(s)|^2 \, ds \le C_2 \sum_{|j| \le l} \|D^j v\|_{loc}^2 \le C_3 \|\eta\|^2,$$

and thus in the space $H(T)$ there exists the weak limit of the derivatives

$$\lim_{n \to \infty} (u_n, \xi^{(k)}) = (u, \xi^{(k)}), \qquad k \le l - 1. \tag{4.20}$$

It is clear that the limit variables $(u, \xi^{(k)})$ belong to the space $H(\Gamma^\delta)$, $\delta > 0$, since $\text{Supp } u_n \subseteq \Gamma^\delta$ for sufficiently large n and consequently $(u, \xi^{(k)}) \in H_+(\Gamma)$. We will show that if Γ is the common boundary of the domains $S_1 = S$ and $S_2 = T \backslash \bar{S}$, then

$$H_+(\Gamma) = \bigvee \{(u, \xi^{(k)}), \text{Supp } u \subseteq \Gamma, k \le l - 1\} \tag{4.21}$$

is the closed linear span of all limit variables (4.20) corresponding to all possible functions of the type (4.19).

We take an arbitrary variable $\eta \in H_+(\Gamma)$ which is orthogonal to all variables $(u, \xi^{(k)})$. For the function $v \in V(T)$ associated with it this means that

$$\int_{\Gamma_{loc}} u(s)\overline{v^{(k)}(s)} \, ds = 0$$

for all $u_{loc} \in C_0^\infty(\Gamma_{loc}^\varepsilon)$, from which it follows that $v^{(k)}(s) = 0$ for all $k \le l - 1$ and almost all $s \subseteq \Gamma_{loc}$ on every piece $\Gamma_{loc} \subseteq \Gamma$. Since the domain $S_1 \cup S_2$ has a bounded compliment in T which coincides with the boundary Γ, it satisfies the duality condition. Therefore under the boundary conditions (4.11) the function $v(t)$, $t \in \Gamma^\varepsilon$, coincides in a neighborhood of the boundary Γ with a function from the class $\mathring{W}_2^l(S_1 \cup S_2)$, from which we conclude—as we did in deriving (4.14)—that $v \in V(S_1 \cup S_2)$. But in the sense of our basic relation (1.9), the space $V(S_1 \cup S_2)$ corresponds to the space $H_+(\Gamma)^\perp$ and $\eta \in H_+(\Gamma)^\perp$; but this is possible for a variable $\eta \in H_+(\Gamma)$ only when $\eta = 0$. Thus the collection of limit variables (4.20) forms a complete system in the space $H_+(\Gamma)$, which we wished to show.

Notice that if the generalized function (4.1) is an "ordinary" random function $\xi(t)$, $t \in T$, having (weak) normal derivatives

$$\frac{\partial^k}{\partial r^k} \xi(t), \qquad k \le l - 1,$$

for almost all $s \in \Gamma$ in a sufficiently small neighborhood of the boundary Γ, then for every variable $\eta \in H(T)$ and associated function $v \in V(T)$ we would have

$$\frac{\partial^k}{\partial r^k} v(t) = \left(\frac{\partial^k}{\partial r^k} \overline{\xi(t)}, \eta \right), \qquad k \le l - 1,$$

and it is evident that the limit variables (4.20) would be

$$(u, \xi^{(k)}) = \int_\Gamma u(s) \frac{\partial^k}{\partial r^k} \overline{\xi(t)} \, ds, \qquad k \le l - 1.$$

Keeping this in mind, we call the limit variables (4.20) the *generalized normal derivatives* (of order k, $k \le l - 1$).

For a generalized function (4.1) and biorthogonal function (4.2) with elliptic form of order $2l$ we have the following result.

Theorem 2. *For domains $S_1 = S$ and $S_2 = T \backslash \bar{S}$ with piecewise l-smooth boundary Γ in common, the boundary space $H_+(\Gamma)$ which splits $H_+(S_1)$ and $H_+(S_2)$ is the closed linear span of the generalized normal derivatives $\xi^{(k)}$, $k \le l - 1$, of the Markov function (4.1) on the boundary Γ.*

Also, as in Theorem 1, under the usual ellipticity condition (4.4) we are referring here to domains S_1 and S_2, at least one of which is bounded; under the stronger condition (4.5) with T_{loc} replaced by T, this additional restriction on S_1, S_2 can be dropped.

4. Regularity and the Dirichlet Problem

Let $T \subseteq \mathbb{R}^d$ be a bounded domain with l-smooth boundary $\Gamma = \partial T$. Each random field $H(S)$, $S \subseteq T$, having a dual field $H^*(S)$, $S \subseteq T$, is *regular* in the sense that

$$H_+(\Gamma) = \bigcap_{\varepsilon > 0} H(\Gamma^\varepsilon) = 0, \qquad (4.22)$$

because

$$[\bigcap H(\Gamma^\varepsilon)]^\perp = [\bigcap H_+(\overline{\Gamma^\varepsilon})]^\perp = \bigvee H_+(\overline{\Gamma^\varepsilon})^\perp = \bigvee H^*(T^{-\varepsilon})$$
$$= H^*(T) = H(T);$$

as usual, Γ^ε denotes an ε-neighborhood of the boundary Γ of the domain T and $T^{-\varepsilon}$ is the complement of the closed set $\overline{\Gamma^\varepsilon}$ in T.

Thus the generalized random function (4.1) we considered in §4.2 and §4.3 is *regular*—i.e., the random field it generates has the regularity property (4.22). Recall that the boundary space $H_+(\Gamma)$, $\Gamma = \partial T$, is the closed linear span of the generalized normal derivatives

$$(u, \xi^{(k)}), \qquad \text{Supp } u \subseteq \Gamma, \qquad k \le l - 1,$$

(cf. (4.21)), which must all be equal to 0 by the regularity condition:

$$(u, \xi^{(k)}) = 0, \qquad k \le l - 1. \qquad (4.23)$$

Let the operator $\mathscr{P}$ in (4.3) have bounded infinitely differentiable co-efficients. Then the random function (4.1) is the solution of the differential equation

$$(\mathscr{P}u, \xi) = (u, \xi^*), \qquad u \in C_0^\infty(T), \tag{4.24}$$

involving an unknown function ξ with generalized biorthogonal function (4.2) given on the right-hand side. To see this, use the function (4.1) for ξ to get

$$(\mathscr{P}u, \xi) = J\mathscr{P}u, \qquad \mathscr{P}u \in C_0^\infty(T),$$

and

$$(J\mathscr{P}u, Lv) = (\mathscr{P}u, v) = (Lu, Lv), \qquad v \in C_0^\infty(T),$$

which gives the equation

$$J\mathscr{P}u = Lu = (u, \xi^*).$$

The generalized random function (4.1) has the property that for any random variables η, $E|\eta|^2 < \infty$, the generalized function which is equal to the scalar product

$$((u, \xi), \eta) = E(u, \xi)\bar\eta = (u, v(\eta)), \qquad u \in C_0^\infty(T),$$

is a function in the class $\mathring{W}_2^l(T)$. We will say that a generalized random function

$$\xi = (u, \xi), \qquad u \in C_0^\infty(T),$$

which has this property *belongs weakly* to the class $\mathring{W}_2^l(T)$.

Recall that the domain T was assumed to be bounded and under condition (4.4) specified above, the generalized functions $v \in V(T)$ belong to the class $\mathring{W}_2^l(T)$. Under the assumption of the uniqueness of the solution to the Dirichlet problem for the homogeneous equation

$$\mathscr{P}v(t) = 0, \qquad t \in T,$$

in the class of generalized functions $v \in \mathring{W}_2^l(T)$ with boundary conditions

$$\frac{\partial^k}{\partial r^k} v(s) = 0, \qquad k \le l - 1,$$

we can say that the generalized random function (4.1) is the unique solution of equation (4.24) with zero boundary conditions (4.23) in the class of all functions which belong weakly to $\mathring{W}_2^l(T)$.

We take

$$\eta = Jv, \qquad v \in C_0^\infty(T),$$

and look at the generalized function

$$B(u, v) = (Ju, \eta) = \int u(t)\overline{B(t, v)} \, dt, \qquad u \in C_0^\infty(T),$$

in the space $V(T)$. Taking the scalar product of both sides of (4.24) with the variable $\eta = Jv$, we get the equation

$$B(\mathscr{P}u, v) = (u, v), \qquad u \in C_0^\infty(T),$$

which can be interpreted for fixed $v \in C_0^\infty(T)$ as the differential equation

$$\mathscr{P}B(t, v) = v(t), \qquad t \in T, \tag{4.25}$$

involving the function $B(\cdot, v) \in \mathring{W}_2^l(T)$, which we know satisfies the boundary conditions

$$\frac{\partial^k}{\partial r^k} B(s, v) = 0, \qquad k \le l - 1, \tag{4.26}$$

for almost all $s \in \Gamma$.

In this way, the covariance function

$$B(u, v) = (Ju, Jv), \qquad u, v \in C_0^\infty(T), \tag{4.27}$$

for the generalized random function (4.1) is the (unique) solution, in the above sense, of equation (4.25) with boundary conditions (4.26).

Earlier we demonstrated a class of Markov generalized functions (4.1) arising from an operator (4.3) which admits a self-adjoint extension $\mathscr{P}$ with bounded inverse operator $\mathscr{P}^{-1}$ and which satisfies a (strong) ellipticity condition in a neighborhood of the closure of the bounded domain $T \subseteq \mathbb{R}^d$— see (3.5) and the discussion following it. In view of the results we have just gotten, we can say that a generalized random function is Markov with respect to all domains $S \subseteq T$ in a bounded domain $T \subseteq \mathbb{R}^d$ if its covariance function satisfies the differential equation (4.25) with the operator $\mathscr{P}$ of the form indicated above and also the boundary conditions (4.26).

Each solution of equation (4.24) differs from the function (4.1) by a term which is a generalized random function satisfying the homogeneous equation

$$(\mathscr{P}u, x) = 0, \qquad u \in C_0^\infty(T). \tag{4.28}$$

The class of "deterministic" functions $v = v(t)$, $t \in T$, connected with this equation is the well-known Sobolev space $W_2^l(T)$, consisting of functions having generalized derivatives $D^k v \in L^2(T)$, $|k| \le l$; this space is similar to $\mathring{W}_2^l(T)$ and differs from it only in that the boundary values $v^{(k)} = (\partial^k/\partial r^k)v(s)$, $k \le l - 1$, on the boundary $\Gamma = \partial T$ of the domain T are not required to be zero.

We consider the class of generalized random functions

$$x = (u, x), \qquad u \in C_0^\infty(T), \tag{4.29}$$

which belong weakly to $W_2^l(T)$ and which have, on the boundary $\Gamma = \partial S$ of any domain $S \subseteq T$ of the type we are considering, weak normal derivatives

$$x^{(k)} = (u, x^{(k)}), \qquad \text{Supp } u \subseteq \Gamma, \qquad k \le l - 1, \tag{4.30}$$

defined in the same way as (4.20). We require that for all $\eta \in H$ and associated $v \in W_2^l(T)$,

$$v = (u, v) = E(u, x)\bar{\eta}, \qquad u \in C_0^\infty(T),$$

and for the normal derivatives $v^{(k)}$, $k \leq l - 1$, defined almost everywhere on $\Gamma = \partial S$ we have

$$(u, v^{(k)}) = E(u, x^{(k)})\bar{\eta}, \qquad \text{Supp } u \subseteq \Gamma.$$

The solution of the homogeneous equation (4.28) on the domain $S \subseteq T$

$$(\mathscr{P}u, x) = 0, \qquad u \in C_0^\infty(S), \tag{4.31}$$

with given boundary values (4.30) is unique in the class of generalized random functions $x = (u, x)$ which belong weakly to $W_2^l(S)$. This is because the difference $x = (u, x)$ between two solutions is a solution in the same class with boundary values zero and for all $\eta \in H$, the associated function $v \in W_2^l(S)$, with $(u, v) = E(u, x)\bar{\eta}$, is a solution of the Dirichlet problem $\mathscr{P}v(t) = 0$, $t \in S$, with zero boundary conditions $v^{(k)}(s) = 0$, $s \in \Gamma$, $k \leq l - 1$, so that $v(t) = 0$ in the domain S and $(u, x) = 0$, $u \in C_0^\infty(S)$.

From the existence of a solution of our Dirichlet problem, we notice that for boundary conditions

$$(u, x^{(k)}) = (u, \xi^{(k)}), \qquad \text{Supp } u \subseteq \Gamma, \qquad k \leq l - 1, \tag{4.32}$$

the solution of equation (4.31) can be described by the following formula:

$$(u, x) = (u, \hat{\xi}) = P_+(\Gamma)(u, \xi), \qquad u \in C_0^\infty(S); \tag{4.33}$$

on the right-hand side the operator $P_+(\Gamma)$ is the projection on the subspace $H_+(\Gamma)$, giving $(u, \hat{\xi})$ as the best approximation to (u, ξ), $u \in C_0^\infty(S)$, under boundary conditions (4.32)—here we are speaking of the Markov random function $\xi = (u, \xi)$ which is a solution of our basic equation (4.24). Indeed, for any variable $\eta \in H_+(\Gamma)$, we get from equation (4.24)

$$E(\mathscr{P}u, \hat{\xi})\bar{\eta} = E(\mathscr{P}u, \xi)\bar{\eta} = E(u, \xi^*)\bar{\eta} = 0, \qquad u \in C_0^\infty(S),$$

since the duality condition implies $H^*(S) \perp H_+(\Gamma)$. It is obvious that the solution (4.33) of equation (4.31) satisfies the boundary conditions (4.32), in which, we recall, the variables $(u, \xi^{(k)}) \in H_+(\Gamma)$ are defined by taking a limit (cf. (4.20)) and

$$\lim(u_n, \xi^{(k)}) = \lim P_+(\Gamma)(u_n, \xi^{(k)}) = P_+(\Gamma)(u, \xi^{(k)}) = (u, \xi^{(k)}).$$

We turn to a generalized random function of the form

$$(u, \xi_x) = (u, \xi) + (u, x), \qquad u \in C_0^\infty(T), \tag{4.34}$$

where $x = (u, x)$ is the solution of the homogeneous equation (4.28) orthogonal to the generalized random function $\xi^* = (u, \xi^*)$ on the right-hand side of our basic equation (4.24):

$$(u, x) \perp H^*(T), \qquad u \in C_0^\infty(T). \tag{4.35}$$

Theorem 3. *Under condition (4.35) the generalized random function (4.34) is Markov†; more precisely, the boundary values $(u, \xi_x^{(k)})$, Supp $u \subseteq \Gamma$, $k \le l - 1$, split the spaces*

$$\bigvee_{\text{Supp}\, u \subseteq S_1} (u, \xi_x) \quad \text{and} \quad \bigvee_{\text{Supp}\, u \subseteq S_2} (u, \xi_x)$$

generated by its values in the domains $S_1 = S$ and $S_2 = T \backslash \bar{S}$ with boundary $\Gamma = \partial S$.

PROOF. Let us take a generalized random function $\xi = (u, \xi)$, $u \in C_0^\infty(S)$, dual to $\xi^* = (u, \xi^*)$, $u \in C_0^\infty(S)$, on the domain $S \subseteq T$. It is a solution of equation (4.24) on the domain S

$$(\mathscr{P}u, \xi) = (u, \xi^*), \qquad u \in C_0^\infty(S),$$

with boundary values zero (cf. (4.23))

$$(u, \xi^{(k)}) = 0, \qquad \text{Supp}\, u \subseteq \Gamma, \qquad k \le l - 1.$$

We take the generalized random function $\overset{\circ}{\xi}_x = (u, \overset{\circ}{\xi}_x)$, $u \in C_0^\infty(S)$, which we get from the projection of the corresponding variables (4.34) on the closed linear span of the boundary values,

$$(u, \xi_x^{(k)}) = (u, \xi^{(k)}) + (u, x^{(k)}), \qquad \text{Supp}\, u \subseteq \Gamma, \qquad k \le l - 1, \quad (4.36)$$

which are orthogonal to the space $H^*(S)$. The function $\overset{\circ}{\xi}_x = (u, \overset{\circ}{\xi}_x)$ is the solution in the domain S of the homogeneous equation (4.31): by the same reasoning as for (4.33) we have

$$E(\mathscr{P}u, \overset{\circ}{\xi}_x)\bar{\eta} = E(\mathscr{P}u, \xi_x)\bar{\eta} = E(u, \xi^*)\bar{\eta} = 0$$

for any variable η from the set of boundary values (4.36). The equation

$$(u, \xi_x) = (u, \xi) + (u, \overset{\circ}{\xi}_x), \qquad u \in C_0^\infty(S), \qquad (4.37)$$

is true since the difference between the left and right sides is a solution of equation (4.31) with zero boundary conditions. The right side of (4.37) is composed of variables $(u, \xi) \in H^*(S)$ and variables $(u, \overset{\circ}{\xi}_x)$ which belong to the closed linear span of the boundary values (4.36). But we know that $H^*(S) = H_+(T \backslash S)^\perp$, and together with condition (4.35) this allows us to conclude that the space of boundary values (4.36) splits the spaces

$$\bigvee_{\text{Supp}\, u \subseteq S_1} (u, \xi_x), \qquad \bigvee_{\text{Supp}\, u \subseteq S_2} (u, \xi_x),$$

consisting of values of the generalized function (4.34) on the complementary domains $S_1 = S$, $S_2 = T \backslash \bar{S}$. $\qquad \square$

Note that from the uniqueness of the "Dirichlet problem" (4.28) for a generalized random function $x = (u, x)$, $u \in C_0^\infty(T)$, with given boundary values (4.30) on the boundary $\Gamma = \partial T$ of the domain T, we get that (cf. (4.33))

$$(u, x) = (u, \hat{x}) = (u, \overset{\circ}{\xi}_x), \qquad u \in C_0^\infty(T),$$

† In connection with the Markov property we should point out that equation (3.15) is true for the corresponding random field.

because together with $x = (u, x)$, the generalized random function $\hat{x} = (u, \hat{x})$ is a solution of equation (4.28) with the same boundary values on $\Gamma = \partial T$ as $x = (u, x)$. Thus the random function $x = (u, x)$ is *singular* and (4.34) gives us a decomposition into regular and singular components.

EXAMPLE (The Brownian Bridge). We consider the random process on the interval $T = (0, 1)$ having the form $\xi(t) = \eta(t) - t\eta(1)$, $0 \leq t \leq 1$, where $\eta(t)$, $t \geq 0$, is the usual Brownian motion with $\eta(0) = 0$. In our general set-up for equation (4.24), the function $\xi(t)$ is associated with the operator $\mathscr{P} = -d^2/dt^2$ and the dual function $\xi^*(t) = -\eta''(t)$; this is immediately apparent from the formulas

$$(u, \xi^*) = \int_0^1 u'(t)\, d\eta(t)$$

$$(u, \xi) = \int_0^1 u(t)\xi(t)\, dt = \int_0^1 [I^2 u(1) - Iu(t)]\, d\eta(t),$$

where $u \in C_0^\infty(0, 1)$ and the operator I is defined as the integral

$$Iu(t) = \int_0^t u(s)\, ds, \qquad 0 \leq t \leq 1.$$

(We note that the boundary values (4.20) for an interval $S = (a, b)$ are simply the values at the boundary points, $\xi(a)$ and $\xi(b)$.) By Theorem 3, for any random linear function $x(t) = \xi_1 + \xi_2 t$, with variables ξ_1, ξ_2 independent of the derivative of "white noise" $\xi^*(t)$, $0 < t < 1$, the random function $\xi_x(t) = x(t) + \xi(t)$, $0 \leq t \leq 1$, is Markov.

Similar results can be obtained in the case of an unbounded domain T with bounded complement by looking at the corresponding "exterior Dirichlet problem."

§5. Stochastic Differential Equations

1. Markov Transformations of "White Noise"

Let

$$(u, \dot{\eta}) = \int u(t)\eta(dt), \qquad u \in C_0^\infty(T)$$

be "white noise" in a domain $T \subseteq \mathbb{R}^d$, generated by the random orthogonal measure $\eta(dt)$,

$$E|\eta(dt)|^2 = dt.$$

As before, we will let $L^2(T)$ be the space of measurable, square-integrable functions $u(t)$, $t \in T$, with scalar product

$$(u, v) = \int u(t)\overline{v(t)}\, dt, \qquad u, v \in L^2(T).$$

Let J be a linear operator in the space $L^2(T)$ with domain $C_0^\infty(T)$ and continuous in the topology of the space $C_0^\infty(T)$ when we consider it as an operator

$$J: C_0^\infty(T) \to H,$$

where H is the closure of the range Ju, $u \in C_0^\infty(T)$, in the space $L^2(T)$. Let

$$(u, \xi) = (Ju, \dot{\eta}), \qquad u \in C_0^\infty(T). \tag{5.1}$$

The question arises: for which operators J will the generalized function (5.1) be Markov?

We will suppose that the adjoint operator J^* has a right inverse L with domain $C_0^\infty(T)$ in the space $L^2(T)$,

$$L: C_0^\infty(T) \to H,$$

and a range which is dense in H, and continuous on the space $C_0^\infty(T)$. This assumption implies the existence of a generalized function biorthogonal to (5.1) and having the form

$$(u, \xi^*) = (Lu, \dot{\eta}), \qquad u \in C_0^\infty(T). \tag{5.2}$$

As we know, under the duality condition for the generalized functions (5.1) and (5.2), the local property of the operator L ensures the Markov property for the random function (5.1) with respect to appropriate domains $S \subseteq T$ — see §1.3 — as well as duality conditions (1.21) and (1.22) and also conditions (3.12) and (3.16) on our operators, which correspond in an obvious way to the operators J and L introduced earlier:

$$Ju = (Ju, \dot{\eta}), \qquad Lu = (Lu, \dot{\eta}).$$

(We rely on it being clear each time which operators we are referring to.)

In the following discussion we will be dealing with a local operator L which is the "square root" of a positive linear differential operator

$$\mathscr{P} = \sum_{|k| \leq 2l} b_k(t)D^k, \tag{5.3}$$

with bounded, infinitely differentiable coefficients $b_k(t)$, $|k| \leq 2l$; more precisely,

$$(Lu, Lv) = (\mathscr{P}u, v), \qquad u, v \in C_0^\infty(T),$$

which implies that the elements Lu, $u \in C_0^\infty(T)$, belong to the domain of the adjoint operator L^* and

$$L^*Lu = \mathscr{P}u, \qquad u \in C_0^\infty(T). \tag{5.4}$$

We note that starting with the differential operator $\mathscr{P}$, one would arrive at a Markov random function (5.1), as we in fact showed in §3.

Let us look at biorthogonal generalized functions (5.1), (5.2). As we have seen (cf. (1.9)) the space $V(T)$ with reproducing kernel for the generalized random function (5.1) is unitarily isomorphic to the space $H(T)$, the closure of all variables $\eta = Ju$, $u \in C_0^\infty(T)$. Using this isomorphism,

$$\eta \leftrightarrow v \in V(T),$$

we get the formula

$$(Lu, \eta) = (\mathscr{P}u, v), \qquad u \in C_0^\infty(T). \tag{5.5}$$

This is obvious for $\eta = Lv$, $v \in C_0^\infty(T)$, and for the case of an arbitrary variable $\eta \in H(T)$, we can take a sequence $\eta_n = Lv_n \to \eta$, where $v_n \to v$ in the space $V(T)$; for every $u \in C_0^\infty(T)$ we have $\mathscr{P}u \in C_0^\infty(T)$ and

$$(Lu, \eta_n) = (\mathscr{P}u, v_n) \to (\mathscr{P}u, v).$$

Recall here that if $v_n \to v$ in the space $V(T)$, then for every $u \in C_0^\infty(T)$,

$$(u, v_n) = (Ju, \eta_n) \to (Ju, \eta) = (u, v).$$

We consider the Markov random field $H(S)$, $S \subseteq T$, generated by the random function (5.1), with its dual random field $H^*(S)$, $S \subseteq T$, generated by the generalized function (5.2) which is biorthogonal to (5.1). Recall that the duality condition means that

$$H_+(S) = H^*(T \backslash \bar{S})^\perp$$

for an appropriate system of domains $S \subseteq T$—say, for all open domains which are bounded or have bounded complements in T.

Each of the spaces $H_+(S) = H_+(\bar{S})$ can be described by making use of the fact that the inclusion $\eta \in H_+(S)$ is equivalent to the variable $\eta \in H(T)$ being orthogonal to all variables Lu, $u \in C_0^\infty(T \backslash \bar{S})$, the closure of which forms $H^*(T \backslash \bar{S})$. By formula (5.5), the variable $\eta \leftrightarrow v$ is orthogonal to the space $H^*(T \backslash \bar{S})$ if and only if

$$(\mathscr{P}u, v) = 0, \qquad u \in C_0^\infty(T \backslash \bar{S}),$$

i.e., when the corresponding function $v \in V(T)$ is the generalized solution of the differential equation

$$\mathscr{P}v(t) = 0, \qquad t \in T \backslash \bar{S}. \tag{5.6}$$

A similar condition results from using the representation

$$\eta = \int Lv(t)\eta(dt), \qquad Lv \in H, \tag{5.7}$$

for variables $\eta \in H(T)$, where the symbol Lv simply indicates a function belonging to the space

$$H = \overline{LC_0^\infty(T)} \subseteq L^2(T).$$

H is the closure of the functions Lu, $u \in C_0^\infty(T)$, where here the symbol Lu can be thought of as the result of applying the operator L to the function u. Namely, let

$$L = \sum_{|k| \le l} a_k(t)D^k \qquad (5.8)$$

be a linear differential operator with bounded, infinitely differentiable coefficients $a_k(t)$, $|k| \le l$, and let

$$L^*u = \sum_{|k| \le l} (-1)^{|k|}D^k[a_k(t)u(t)], \qquad u \in C_0^\infty(T), \qquad (5.9)$$

be the linear differential operator which is formally adjoint to L; notice that

$$L^*Lu = \mathcal{P}u, \qquad u \in C_0^\infty(T).$$

Then the condition that the variable (5.7) be orthogonal to the space $H^*(T \setminus \bar{S})$ implies that the function $Lv \in \overline{LC_0^\infty(T)}$ is the generalized solution of the differential equation

$$L^*(Lv(t)) = 0, \qquad t \in T \setminus \bar{S}. \qquad (5.10)$$

Of course, under conditions (5.6) and (5.10), one can take, instead of a closed domain $\bar{S}$, any closed set Γ in T for which the domain $T \setminus \Gamma$ satisfies the duality condition

$$H_+(\Gamma) = H^*(T \setminus \Gamma)^\perp,$$

(we assume that Γ is a bounded set). In particular, for the boundary $\Gamma = \partial S$ of any domain $S \subseteq T$ we have the following description of the boundary space $H_+(\Gamma)$. A variable $\eta \in H(T)$ belongs to the space $H_+(\Gamma)$ if and only if the associated function $v \in V(T)$ is a generalized solution of the equation

$$\mathcal{P}v(t) = 0, \qquad t \in T \setminus \Gamma, \qquad (5.11)$$

or correspondingly, the function $Lv \in H$ is a generalized solution of the equation

$$L^*(Lv(t)) = 0, \qquad t \in T \setminus \Gamma. \qquad (5.12)$$

Let L be an operator of the form (5.8) with range space dense in $L^2(T)$; this occurs for a quite wide class of differential operators in the case where the domain $T \subseteq \mathbb{R}^d$ is unbounded. Then

$$H = \overline{LC_0^\infty(T)} = L^2(T)$$

and the adjoint operator L^* is defined on the space $C_0^\infty(T)$, where it coincides with the formal adjoint of the differential operator (5.9), moreover

$$L^*u \in C_0^\infty(T), \qquad u \in C_0^\infty(T).$$

The inverse operator $(L^*)^{-1}$ exists because if $L^*v = 0$ then

$$(L^*v, u) = (v, Lu) = 0$$

for all $u \in C_0^\infty(T)$, hence $v = 0$.

It is obvious that the operator J in (5.1), being a right inverse of the adjoint operator L^*,

$$L^*Ju = u, \qquad u \in C_0^\infty(T), \tag{5.13}$$

is a restriction of the operator $J = (L^*)^{-1}$. In fact, applying the inverse operator $(L^*)^{-1}$ to both sides of (5.13) we get

$$Ju = (L^*)^{-1}u, \qquad u \in C_0^\infty(T).$$

Moreover, starting with an operator L of the indicated type we can say that a generalized function (5.2) has a biorthogonal function if and only if the space $C_0^\infty(T)$ is contained in the domain of the inverse operator $(L^*)^{-1}$; in this case the biorthogonal function (5.1) is given by the operator $J = (L^*)^{-1}$ on the space $C_0^\infty(T)$ satisfying not only condition (5.13), but also

$$JL^*u = u, \qquad u \in C_0^\infty(T). \tag{5.14}$$

Equation (5.14) shows that the random function (5.1) is a generalized solution of the stochastic differential equation

$$L\xi(t) = \dot{\eta}(t), \qquad t \in T, \tag{5.15}$$

with "white noise" on the right-hand side; more precisely,

$$(u, L\xi) = (L^*u, \xi) = (JL^*u, \dot{\eta}) = (u, \dot{\eta}), \qquad u \in C_0^\infty(T).$$

We call the generalized function (5.1) a *fundamental solution* of the stochastic equation (5.15). In this connection we note that for a wide class of linear differential operators L, the operator $J = (L^*)^{-1}$ is given by a kernel $J(t, s)$ which can be represented as a so-called fundamental solution of a deterministic differential equation, namely

$$L^*J(t, s) = \delta(t - s),$$

where $\delta(\cdot)$ denotes the usual δ-function.

We summarize the results above in the form of a separate proposition for a linear differential operator L with a range space which is dense in $L^2(T)$ and having the property that

$$\|L(w \cdot u)\| \le C(w)\|Lu\|, \qquad u \in C_0^\infty(T), \tag{5.16}$$

for a function $w \in C_0^\infty(\mathbb{R}^d)$. See, in connection with this, (3.16) and (3.17).

Theorem. *A fundamental solution of the stochastic differential equation* (5.15) *has the Markov property with respect to all open domains $S \subseteq T$ which are bounded or have bounded complements in T. Moreover, the boundary space $H_+(\Gamma)$, $\Gamma = \partial S$, splitting $H_+(S)$ and $H_+(T \setminus \bar{S})$ consists of variables having the following form*:

$$\eta = (u, \dot{\eta}) = \int u(t)\eta(dt), \tag{5.17}$$

where $u \in L^2(T)$ is a generalized solution of the differential equation
$$L^*u(t) = 0, \qquad t \in T \setminus \Gamma.$$

2. The Interpolation and Extrapolation Problems

We will consider the problem of making a best estimate of the random function (5.1) based on its values in a domain $S \subseteq T$; more precisely, we look at the problem of finding a best approximation to a variable $\eta \in H_+(T \backslash \bar{S})$ by variables in the space $H_+(S)$—i.e., finding a variable $\hat{\eta} \in H_+(S)$ such that $\hat{\eta}$ minimizes the mean square error among all possible approximations in $H_+(S)$:

$$\| \eta - \hat{\eta} \| = \min.$$

In the case where we distinguish "interior" and "exterior" domains $S_1 \subseteq T$ and $S_2 = T \backslash \bar{S}_1$, the problem of finding the best approximation is called an *extrapolation problem* if S is the interior domain, $S = S_1$, and an *interpolation problem* if S is the exterior region, $S = S_2$.

We denote by $P_+(S)$ the orthogonal projection operator on the space $H_+(S)$. The best approximation for a variable η is given by its projection $\hat{\eta} = P_+(S)\eta$ on $H_+(S)$.

We suppose that there exists a biorthogonal generalized function (5.2) generating the dual random field

$$H^*(S) = H_+(S^c)^\perp, \qquad S \subseteq T.$$

Let the variable $\eta \in H_+(T \backslash \bar{S})$ be given by the stochastic integral (5.7)

$$\eta = \int Lv(t)\eta(dt), \tag{5.18}$$

where the related function $Lv \in \overline{LC_0^\infty(T)}$ and is, by (5.10), a generalized solution of the equation

$$L^*Lv(t) = 0, \qquad t \in S.$$

For a Markov random function (5.1) we have

$$P_+(S)H_+(T \backslash \bar{S}) = H_+(\Gamma),$$

where $H_+(\Gamma)$, $\Gamma = \partial S$, is the boundary space splitting $H_+(S)$ and $H_+(T \backslash \bar{S})$, so that

$$\hat{\eta} = P_+(S)\eta \in H_+(\Gamma)$$

and we can represent $\hat{\eta}$ by the stochastic integral

$$\hat{\eta} = \int L\hat{v}(t)\eta(dt), \tag{5.19}$$

where the function $L\hat{v}(t) \in \overline{LC_0^\infty(T)}$ must be a generalized solution of

$$L^*L\hat{v}(t) = 0, \qquad t \in T \backslash \Gamma,$$

cf. (5.12). The difference $\hat{\eta} - \eta$ is uniquely defined by the conditions

$$\hat{\eta} \in H_+(\Gamma), \qquad \hat{\eta} - \eta \in H_+(S)^\perp = H^*(T \backslash \bar{S});$$

the second of these is equivalent to the inclusion

$$L\hat{v} - Lv \in \overline{LC_0^\infty(T\setminus\bar{S})}, \tag{5.20}$$

because $H^*(T\setminus\bar{S})$ is the closure of all variables

$$\eta = \int Lu(t)\eta(dt), \qquad u \in C_0^\infty(T\setminus\bar{S}).$$

It is obvious that condition (5.20) includes the equation

$$L\hat{v}(t) = Lv(t), \qquad t \in S, \tag{5.21}$$

and the best approximation problem leads to finding, in the domain $T\setminus\bar{S}$, a solution to

$$L^*L\hat{v}(t) = 0, \qquad t \in T\setminus\bar{S}, \tag{5.22}$$

satisfying condition (5.20). Indeed, such a solution, together with the function $Lv(t)$ solving equation (5.22) on the domain S, gives the variable (5.19) in the boundary space $H_+(\Gamma)$; furthermore, (5.20) implies that the difference $\hat{\eta} - \eta$ belongs to the space $H^*(T\setminus\bar{S}) = H_+(S)^\perp$.

Suppose that instead of a function $Lv \in \overline{LC_0^\infty(T)}$ we are given a function $v \in V(T)$ corresponding to the variable $\eta \leftrightarrow v$ via the unitary isomorphism (1.9). Notice that if the operator L has a bounded inverse, the function $v \leftrightarrow \eta$ belongs to the domain $D(T) \subseteq L^2(T)$ of the closure of L, and the function Lv above is the result of applying the (closed) operator L to the function $v \in D(T)$, $v = L^{-1}(Lv)$—see (3.5)–(3.10).

From the general condition (5.6) we have for a function

$$v \leftrightarrow \eta \in H_+(T\setminus\bar{S}), \tag{5.23}$$

the differential equation

$$\mathscr{P}v(t) = 0, \qquad t \in S,$$

and for the corresponding function

$$\hat{v} \leftrightarrow \hat{\eta} \in H_+(\Gamma), \tag{5.24}$$

the equation

$$\mathscr{P}\hat{v}(t) = 0, \qquad t \in T\setminus\Gamma$$

cf. (5.11). Clearly, the function $\hat{v} \in V(T)$ is uniquely defined by the fact that it satisfies the differential equation

$$\mathscr{P}\hat{v}(t) = 0, \qquad t \in T\setminus\bar{S}, \tag{5.25}$$

and the condition

$$\hat{v} - v \in V(T\setminus\bar{S}); \tag{5.26}$$

recall that

$$V(T\setminus\bar{S}) \leftrightarrow H^*(T\setminus\bar{S}) = H_+(S)^\perp.$$

Notice that for an elliptic operator $\mathscr{P} = L*L$ of order $2l$ and piecewise l-smooth boundary $\Gamma = \partial S$, condition (5.26) can be replaced by the Dirichlet boundary conditions on the normal derivatives at the boundary Γ:

$$\frac{\partial^k}{\partial r^k} \hat{v}(s) = \frac{\partial^k}{\partial r^k} v(s), \qquad k \leq l - 1. \qquad (5.27)$$

Recall that the functions $\hat{v}$, $v \in V(T)$, belong to the class $\overset{\circ}{W}{}^l_2(T)$ for all domains T if (4.5) holds or belong locally to $\overset{\circ}{W}{}^l_2(T)$ for bounded domains $T_{\text{loc}} \subseteq T$ if condition (4.4) holds; in the latter case our results relate to domains $S \subseteq T$ which are bounded or have bounded complements in T. We will elaborate on what we mean by replacing (5.26) by the boundary conditions (5.27). Let $u = \hat{v} - v$. Take a function $w \in C_0^\infty(\mathbb{R}^d)$ with $w(t) = 1$ for $t \in \Gamma^\varepsilon$. The conditions (5.27) imply that

$$u \cdot w \in \overset{\circ}{W}{}^l_2(T\backslash\Gamma) \subseteq V(T\backslash\Gamma).$$

Since $u - u \cdot w = 0$ on the domain Γ^ε, we have

$$(u - u \cdot w) \in V(T\backslash\Gamma), \qquad u = u \cdot w + (u - u \cdot w) \in V(T\backslash\Gamma).$$

Clearly the function

$$\hat{u} = \begin{cases} u, & \text{on the domain } T\backslash\bar{S}, \\ 0, & \text{on the domain } \bar{S}, \end{cases}$$

belongs to the space $V(T)$, together with the function

$$\hat{v} = \hat{u} + v = \begin{cases} \hat{v}, & \text{on } T\backslash\bar{S}, \\ v, & \text{on } \bar{S}, \end{cases}$$

where the $\hat{v}$ appearing on the right side is a solution of equation (5.25) in the domain $T\backslash\bar{S}$ with boundary conditions (5.27).

3. The Brownian Sheet

It will be convenient for us to denote the standard Gaussian orthogonal measure by the symbol $d\xi(t)$, since we will be considering here a Gaussian random function of the form

$$\xi(t) = \int_0^t d\xi(s), \qquad t \geq 0,$$

where $t \geq 0$ means that every coordinate of the point $t \in \mathbb{R}^d$ is ≥ 0 and for each $t \geq 0$ we are integrating over the domain $0 \leq s \leq t$. This function $\xi(t)$, $t \geq 0$, with multi-dimensional parameter $t \in \mathbb{R}^d$, is a generalization of the well-known Wiener process (the Brownian motion process); in the case of a two-dimensional parameter $t = (t_1, t_2)$, this is usually called the *Brownian sheet*.

It is elementary to verify that the Brownian sheet has the Markov property with respect to domains S which are composed of rectangles with sides parallel to the coordinate axes. Namely, for each point t it is easy to show a linear combination of values $\xi(s)$, $s \in \Gamma$, on the boundary $\Gamma = \partial S$ which represents the projection of the variable $\xi(t)$ on the space $H(\bar{S})$—the closure of the values of our function in the domain $\bar{S}$. This can be done, for example, by successively looking at rectangles (s, t) and using the representation

$$\xi(t_1, t_2) - [\xi(s_1, t_2) + \xi(t_1, s_2) - \xi(s_1, s_2)] = \int_s^t d\xi(u).$$

For instance, for the rectangle (s, t) shown in Figure 1 the bracket on the left-hand side contains values on the boundary Γ of the domain S and the stochastic integral on the right is a variable orthogonal to the space $H(\bar{S})$.

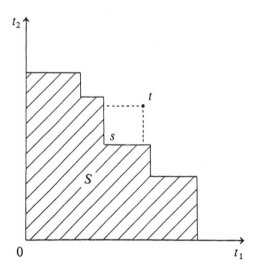

Figure 1

One might imagine that the property shown above for the Brownian sheet, namely

$$P(H(\bar{S}))H(T \backslash \bar{S}) = H(\Gamma),$$

where $H(\Gamma)$ is the span of the variables $\xi(t)$, $t \in \Gamma$, could be extended by a suitable limit procedure to domains with sufficiently "nice" boundaries of some general type. But this is not so. This property does not hold even for a domain as simple as a triangle; we will demonstrate this somewhat later. (However, at the same time, the Brownian sheet has the Markov property with respect to all domains $S \subseteq T$ which are bounded or have bounded complements in the domain $T = \{t > 0\}$.)

We will show that the Brownian sheet, regarded as a generalized random function, is a fundamental solution of the stochastic equation (5.15) with differential operator

$$L = \frac{\partial^2}{\partial t_1 \, \partial t_2}$$

and associated "white noise" $\dot{\eta}(t) = \dot{\xi}(t)$,

$$(u, \dot{\xi}) = \int u(t) \, d\xi(t), \qquad u \in C_0^\infty(T),$$

on the domain $T = \{t = (t_1, t_2) > 0\}$.

We consider the kernel

$$J(s, t) = \begin{cases} 1, & s \in (0, t), \\ 0, & s \notin (0, t), \end{cases} \qquad s, t \in T,$$

and the operator

$$Ju(t) = \int J(t, s)u(s) \, ds = \int_t^\infty u(s) \, ds, \qquad u \in C_0^\infty(T).$$

We have

$$J^*Lu(t) = \int_0^t Lu(s) \, ds = \int_0^{t_1} \int_0^{t_2} \frac{\partial^2}{\partial s_1 \, \partial s_2} u(s) \, ds_1 \, ds_2$$

$$= u(t), \qquad t = (t_1, t_2),$$

and moreover,

$$JL^*u(t) = \int_t^\infty L^*u(s) \, ds = \int_{t_1}^\infty \int_{t_2}^\infty \frac{\partial^2}{\partial s_1 \, \partial s_2} u(s) \, ds_1 \, ds_2$$

$$= \int_{t_1}^\infty \left[-\frac{\partial}{\partial s_1} u(s_1, t_2) \right] ds_1 = u(t)$$

for all $u \in C_0^\infty(T)$; that is, conditions (5.13) and (5.14) are satisfied. Finally, we note that

$$(u, \xi) = \int u(t)\xi(t) \, dt = \iint u(t)J(s, t) \, d\xi(s) \, dt$$

$$= \iint u(s)J(t, s) \, d\xi(t) \, ds = \int \left[\int J(t, s)u(s) \, ds \right] d\xi(t) = (Ju, \dot{\xi}),$$

$$u \in C_0^\infty(T).$$

We will check that the hypotheses of our theorem about the Markov property for fundamental solutions of stochastic differential equations of the form (5.15) are satisfied.

One of these hypotheses was the completeness of the range $Lu, u \in C_0^\infty(T)$, in the space $L^2(T)$. It is clear that one can approximate in $L^2(T)$ the function $v(t) = v_1(t_1) \cdot v_2(t_2)$ by functions

$$\frac{\partial^2}{\partial t_1 \, \partial t_2} u(t) = u_1'(t_1) \cdot u_2'(t_2),$$

where $u(t) = u_1(t_1) \cdot u_2(t_2) \in C_0^\infty(T)$, provided the factors $v_1(t_1)$, $v_2(t_2)$ have compact support and satisfy

$$\int_0^\infty v_1(t_1) \, dt_1 = \int_0^\infty v_2(t_2) \, dt_2 = 0.$$

It is also clear that since the measure of the domain $T = \{t > 0\}$ is infinite, the functions above can approximate in $L^2(T)$ any function of the form $v(t) = v_1(t_1) \cdot v_2(t_2)$, hence they form a complete system in $L^2(T)$.

It remains to verify that our operator $L = \partial^2/\partial t_1 \, \partial t_2$ satisfies (5.16) in the domain $T = \{t > 0\}$. Using the Fourier transform

$$\tilde{u}(\lambda) = \int \exp(i\lambda t)u(t) \, dt, \qquad \lambda = (\lambda_1, \lambda_2),$$

we represent functions $u \in C_0^\infty(T)$ in the form

$$u(t) = \frac{1}{(2\pi)^2} \int \frac{\exp(-i\lambda_1 t_1) - 1}{-i\lambda_1} \cdot \frac{\exp(-i\lambda_2 t_2) - 1}{-i\lambda_2} \varphi(\lambda) \, d\lambda,$$

where

$$\varphi(\lambda) = \widetilde{Lu}(\lambda) = (-i\lambda_1)(-i\lambda_2)\tilde{u}(\lambda);$$

for any bounded domain $T_{\text{loc}} \subseteq T$ we easily get estimates

$$\|u\|_{\text{loc}} \le C_0 \|\varphi\|; \quad \left\| \frac{\partial u}{\partial t_1} \right\|_{\text{loc}}, \left\| \frac{\partial u}{\partial t_2} \right\|_{\text{loc}} \le C_1 \|\varphi\|;$$

$$\left\| \frac{\partial^2 u}{\partial t_1 \, \partial t_2} \right\|_{\text{loc}} \le C_2 \|\varphi\|.$$

We will explicitly derive only one of them. Say,

$$\frac{\partial}{\partial t_1} u(t) = \frac{1}{(2\pi)^2} \int \exp(-i\lambda_1 t_1) \left[\int \frac{\exp(-i\lambda_2 t_2) - 1}{-i\lambda_2} \varphi(\lambda_1, \lambda_2) \, d\lambda_2 \right] d\lambda_1,$$

where, for fixed t_2, the integral in brackets on the right is a square-integrable function of λ_1; then

$$\int \left| \frac{\partial}{\partial t_1} u(t) \right|^2 dt_1 \le C \iint \left| \frac{\exp(-i\lambda_2 t_2) - 1}{-i\lambda_2} \right|^2 |\varphi(\lambda_1, \lambda_2)|^2 d\lambda_1 \, d\lambda_2$$

$$\left\| \frac{\partial}{\partial t_1} u \right\|_{\text{loc}} \le C_1 \cdot \|\varphi\|.$$

From our bounds for the domain $T_{\text{loc}} \supseteq T \cap \operatorname{Supp} w$, it follows that

$$\|L(u \cdot w)\| \leq C\left(\|u\|_{\text{loc}} + \left\|\frac{\partial u}{\partial t_1}\right\|_{\text{loc}} + \left\|\frac{\partial u}{\partial t_2}\right\|_{\text{loc}} + \left\|\frac{\partial^2 u}{\partial t_1 \partial t_2}\right\|_{\text{loc}}\right)$$

$$\leq C(w) \cdot \|Lu\|.$$

In short, we get that the Brownian sheet has the Markov property with respect to all domains $S \subseteq T$ which are bounded or have bounded complements in the domain $T = \{t > 0\}$. Moreover, the boundary space $H_+(\Gamma)$, $\Gamma = \partial S$, which splits the spaces $H_+(S)$ and $H_+(T\backslash \bar{S})$, consists of all variables of the form

$$\eta = \int u(t) \, d\xi(t),$$

where $u \in L^2(T)$ is the generalized solution of the equation

$$\frac{\partial^2}{\partial t_1 \partial t_2} u(t) = 0, \qquad t \in T\backslash\Gamma,$$

cf. (5.17).

We will show that the boundary space $H_+(\Gamma)$ can be richer than just the closed linear span of the boundary values $\xi(t)$, $t \in T$. As an example we will look at the triangular domain S under the segment $\Gamma = \{t_1 + t_2 = 1\}$. We take a (nonconstant) function $u_1(t_1)$ and we let

$$f(t_1) = \frac{1}{t_1} \int_0^{t_1} u_1(t_1) \, dt_1, \qquad u_2(t_2) = \frac{d}{dt_2} t_2 f(1 - t_2).$$

Then

$$\frac{1}{t_2} \int_0^{t_2} u_2(t_2) \, dt_2 = \frac{1}{t_1} \int_0^{t_1} u_1(t_1) \, dt_1, \qquad t = (t_1, t_2) \in \Gamma,$$

and the (not identically zero) function

$$u(t) = \begin{cases} u_1(t_1) - u_2(t_2), & t \in S_1 = S, \\ 0, & t \in S_2 = T\backslash\bar{S}, \end{cases}$$

is such that the variable

$$\eta = \int u(t) \, d\xi(t) \in H_+(\Gamma)$$

is independent of the variables $\{\xi(t), t \in \Gamma\}$, since

$$(\eta, \xi(t)) = \int_0^{(t_1, t_2)} u(s) \, ds = 0, \qquad t \in \Gamma.$$

If we show that η belongs to both the space $H(S_1)$ and the space $H(S_2)$ then it will be obvious that $H(S_1)$ and $H(S_2)$ are not conditionally independent with respect to $H(\Gamma)$, since otherwise the variable $\eta \neq 0$, being independent of $H(\Gamma)$, would be independent of itself. Under the similarity transformation $t \rightsquigarrow \alpha t$, with coefficient α near 1, the function $u(t) = u_1(t_1) - u_2(t_2)$ maps into the function $u(\alpha t)$, differing only slightly from $u(t)$ and

$$\int |u(t) - u(\alpha t)|^2 \, dt \to 0 \quad \text{as } \alpha \to 1.$$

The approximating function $u(\alpha t)$ satisfies the condition

$$\frac{\partial^2}{\partial t_1 \, \partial t_2} u(\alpha t) = 0, \qquad t \notin \Gamma_\alpha = \{t_1 + t_2 = 1/\alpha\},$$

and therefore

$$\eta_\alpha = \int u(\alpha t) \, d\xi(t) \in H_+(\Gamma_\alpha),$$

where Γ_α lies strictly inside S_1 for $\alpha > 1$ and strictly inside S_2 for $\alpha < 1$, so that $\eta_\alpha \in H(S_1)$ in the first case and $\eta_\alpha \in H(S_2)$ in the second; clearly

$$\eta = \lim_{\alpha \to 1} \eta_\alpha \in H(S_1) \cap H(S_2).$$

Note that we have also shown that $H(\Gamma)$ does not split $H(S_1)$ and $H(S_2)$.

CHAPTER 4
Vector-Valued Stationary Functions

§1. Conditions for Existence of the Dual Field

1. Spectral Properties

We will consider stationary random functions $\xi(t)$, $t \in \mathbb{R}^d$, whose values are random elements (vectors) in a Hilbert space X with norm $\|\xi(t)\|$,

$$E\|\xi(t)\|^2 < \infty, \qquad t \in \mathbb{R}^d, \tag{1.1}$$

and also random fields formed by the spaces $H(S)$, $S \subseteq \mathbb{R}^d$, each of which is the closed linear span of the variables in the space $L^2(\Omega, \mathscr{A}, P)$

$$\xi_x(t) = (\xi(t), x) \tag{1.2}$$

defined by the scalar product in X of $\xi(t)$, $t \in S$, and $x \in X$. *Stationarity* will be understood in the following sense:

$$\xi_x(t + s) = U_s \xi_x(t), \qquad t, s \in \mathbb{R}^d, \tag{1.3}$$

for all $x \in X$, where U_s, $s \in \mathbb{R}^d$, form a continuous group of unitary operators in the space $H(\mathbb{R}^d)$, usually called the *shift operators*.

We will suppose that a stationary function $\xi(t)$, $t \in \mathbb{R}^d$, with values in a Hilbert space X, has expectation zero and spectral density $f(\lambda)$—i.e., a positive operator-valued function of $\lambda \in \mathbb{R}^d$ such that

$$(\xi_x(t), \xi_y(s)) = (U_{t-s}\xi_x(0), \xi_y(0))$$

$$= \int \exp[i\lambda(t - s)](f(\lambda)x, y) \, d\lambda, \qquad x, y \in X, t, s \in \mathbb{R}^d \tag{1.4}$$

Its Fourier transform gives us the (operator-valued) *covariance function*

$$B(t) = \int_{\mathbb{R}^d} \exp(i\lambda t) f(\lambda) \, d\lambda, \qquad t \in \mathbb{R}^d.$$

We know that under condition (1.1), the covariance operator $B = B(0)$, common to all the random vectors $\xi(t) \in X$, is nuclear, that is,

$$\sum_k (Bx_k, x_k) = \int_{\mathbb{R}^d} \sum_k (f(\lambda)x_k, x_k) \, d\lambda < \infty,$$

where $\{x_k\}$ is an orthonormal basis for X. Thus the spectral density $f(\lambda)$ is a nuclear operator for a.e. λ and its trace class norm

$$\|f(\lambda)\|_1 = \sum_k (f(\lambda)x_k, x_k)$$

is integrable.

It will be convenient to consider the Hilbert space $L^2(\mathbb{R}^d, X)$ of measurable functions $x(\lambda)$, $\lambda \in \mathbb{R}^d$, with values in X, having square-integrable norm $\|x(\lambda)\|$ and scalar product

$$\{x(\lambda), y(\lambda)\} = \int (x(\lambda), y(\lambda)) \, d\lambda, \qquad x(\lambda), y(\lambda) \in L^2(\mathbb{R}^d, X).$$

In particular, along with the family of variables

$$\xi_x(t), \qquad x \in X, t \in \mathbb{R}^d,$$

we will consider the unitarily isomorphic family of functions

$$x(\cdot, t) = \exp(i\lambda t) f^{1/2}(\lambda)x, \qquad x \in X, t \in \mathbb{R}^d,$$

and corresponding to the indicated functions $x(\cdot, t)$, $t \in S$, the subspaces

$$H(S) = \bigvee_{t \in S} \exp(i\lambda t) f^{1/2}(\lambda)X, \qquad S \subseteq \mathbb{R}^d, \tag{1.5}$$

where $f^{1/2}(\lambda)$ is the square root of the positive operator-valued function $f(\lambda)$.

We introduce a subspace-valued function $\overline{A(\lambda)}$, $\lambda \in \mathbb{R}^d$, for any $A \subseteq L^2(\mathbb{R}^d, X)$, by choosing a complete set of functions $\{a_1(\lambda), a_2(\lambda), \ldots\}$ in A and defining $\overline{A(\lambda)}$ for almost all λ as the closed linear span in X of the values $a_1(\lambda), a_2(\lambda), \ldots$. Note that for any function $a(\lambda) \in A$ there is sequence of linear combinations $\sum_k c_k a_k(\lambda)$ which converges in $L^2(\mathbb{R}^d, X)$ to $a(\lambda)$ and thus some subsequence $\sum_k c_k a_k(\lambda)$ converges almost everywhere to $a(\lambda)$, $a(\lambda) \in \overline{A(\lambda)}$ for almost all λ, and for any other complete set $\{a_1'(\lambda), a_2'(\lambda), \ldots\}$ in $A \subseteq L^2(\mathbb{R}^d, X)$, the corresponding subspace-valued function $\overline{A'(\lambda)}$ will be the same as $\overline{A(\lambda)}$ for almost all λ. In this sense, the subspace-valued function $\overline{A(\lambda)}$, $\lambda \in \mathbb{R}^d$, is uniquely defined for almost all λ.

Lemma. *The subspace*

$$H = \bigvee_{t \in \mathbb{R}^d} \exp(i\lambda t)A \subseteq L^2(\mathbb{R}^d, X)$$

consists of all functions $x(\lambda)$, $\lambda \in \mathbb{R}^d$, in $L^2(\mathbb{R}^d, X)$ whose values satisfy the condition

$$x(\lambda) \in \overline{A(\lambda)} \quad \text{a.e.} \tag{1.6}$$

PROOF. By definition, each function $x(\lambda) \in H$ is the limit of functions of the form $\sum_k \exp(i\lambda t_k) a_k(\lambda)$, where $a_k(\lambda) \in A$; clearly values of $x(\lambda)$ satisfy condition (1.6). Conversely, notice that for any function $a(\lambda) \in A$ and scalar measurable function $c(\lambda)$ with

$$\int |c(\lambda)|^2 \, \|a(\lambda)\|^2 \, d\lambda < \infty,$$

the product $x(\lambda) = c(\lambda)a(\lambda)$ is a function in the subspace H because the function $c(\lambda)$ can be arbitrarily closely approximated in mean square (with weighting $g(\lambda) = \|a(\lambda)\|^2$) by linear combinations of the form $\sum_k c_k \exp(i\lambda t_k)$ and for some sequence of the form $\sum_k c_k \exp(i\lambda t_k) a(\lambda)$ in $L^2(\mathbb{R}^d, X)$ we have

$$\int \left\| x(\lambda) - \sum_k c_k \exp(i\lambda t_k) a(\lambda) \right\|^2 d\lambda = \int \left| c(\lambda) - \sum_k c_k \exp(i\lambda t_k) \right|^2 g(\lambda) \, d\lambda \to 0.$$

Similarly, for any functions $a_1(\lambda), \ldots, a_r(\lambda) \in A$ and measurable vector function $c(\lambda) = (c_1(\lambda), \ldots, c_r(\lambda))$ satisfying the condition

$$\int c(\lambda)g(\lambda)c(\lambda)^* \, d\lambda < \infty,$$

with $g(\lambda)$ a positive matrix-valued function having components

$$g_{pq}(\lambda) = \{a_p(\lambda), a_q(\lambda)\}, \qquad p, q = 1, \ldots, r,$$

the function $x(\lambda) = \sum_{p=1}^r c_p(\lambda)a_p(\lambda)$ belongs to the subspace H; this follows because the vector function $c(\lambda)$ can be arbitrarily closely approximated in mean square (with matrix weighting $g(\lambda)$) by linear combinations of the form $\sum_k c_k \exp(i\lambda t_k)$ with vector components $c_k = (c_{k1}, \ldots, c_{kr})$, and for some sequence of functions $\sum_k \exp(i\lambda t_k) \sum_{p=1}^r c_{kp} a_p(\lambda)$ in $L^2(\mathbb{R}^d, X)$ we have

$$\int \left\| x(\lambda) - \sum_k \exp(i\lambda t_k) \sum_{p=1}^r c_{kp} a_p(\lambda) \right\|^2 d\lambda$$

$$= \int \left[c(\lambda) - \sum_k c_k \exp(i\lambda t_k) \right] g(\lambda) \left[c(\lambda) - \sum_k c_k \exp(i\lambda t_k) \right]^* d\lambda \to 0.$$

Now take an arbitrary function $x(\lambda) \in L^2(\mathbb{R}^d, X)$ satisfying (1.6) and take a complete set of functions $a_1(\lambda), a_2(\lambda), \ldots \in A$ whose values generate the subspace $\overline{A(\lambda)} \subseteq X$ for almost all $\lambda \in \mathbb{R}^d$. The projection in the Hilbert space X of the variable $x(\lambda) \in \overline{A(\lambda)}$ on the subspace generated by a finite number of variables $a_1(\lambda), \ldots, a_r(\lambda)$ has the form $x_r(\lambda) = \sum_{p=1}^r c_p(\lambda)a_p(\lambda)$; as functions of λ the coefficients $c_1(\lambda), \ldots, c_r(\lambda)$ are measurable by virtue of the fact that for any functions $x(\lambda), y(\lambda) \in L^2(\mathbb{R}^d, X)$, the scalar product $\{x(\lambda), y(\lambda)\}$ is a measurable function of λ. We showed before that $x_r(\lambda) \in H$. But for every fixed λ

$$\|x(\lambda) - x_r(\lambda)\|^2 \to 0 \quad \text{as } r \to \infty,$$

and

$$\|x(\lambda) - x_r(\lambda)\|^2 \leq \|x(\lambda)\|^2,$$

so

$$\int \|x(\lambda) - x_r(\lambda)\|^2 \, d\lambda \to 0 \quad \text{as } r \to \infty;$$

this gives $x(\lambda) \in H$ and the lemma is proved. $\square$

According to this lemma, the subspace

$$H(\mathbb{R}^d) = \bigvee_{t \in \mathbb{R}^d} \exp(i\lambda t) f^{1/2}(\lambda) X \subseteq L^2(\mathbb{R}^d, X)$$

consists of all those functions $x(\lambda) \in L^2(\mathbb{R}^d, X)$ whose values satisfy the condition

$$x(\lambda) \in \overline{f^{1/2}(\lambda) X} \quad \text{a.e.} \tag{1.7}$$

2. Duality

We consider the family

$$H^*(S) = H_+(S^c)^\perp, \qquad S^c = \mathbb{R}^d \backslash S, \tag{1.8}$$

where $H(S)$, $S \subseteq \mathbb{R}^d$, is a stationary field (1.5) which is unitarily isomorphic to the random field generated by the stationary random functions $\xi(t)$, $t \in \mathbb{R}^d$, i.e., to the random field associated with the closed linear spans of the variables (1.2), with parameters $x \in X$, $t \in S$. We will consider $H^*(S)$ over the system $\mathscr{G}$ of all open domains $S \subseteq \mathbb{R}^d$ which are bounded or have bounded complements and we will show that the family $H^*(S)$, $S \in \mathscr{G}$, is additive over this system:

$$H^*(S' \cup S'') = H^*(S') \vee H^*(S'')$$

for S', $S'' \in \mathscr{G}$.

We call the family $H^*(S)$, $S \in \mathscr{G}$, the *dual (or conjugate)* field of $H(S)$, $S \in \mathscr{G}$. One can verify that it plays the same role in with regard to the Markov property as the dual field we considered previously (see Chapter 2, §3.5). The only difference is that now our fields are being considered on domains $S \in \mathscr{G}$.

Let $H(S)^\perp$ be the orthogonal complement of the subspace

$$H(S) = \bigvee_{t \in S} \exp(i\lambda t) f^{1/2}(\lambda) X$$

in the space $H(\mathbb{R}^d)$. If $a(\lambda) \in H(S)^\perp$ then

$$\int (a(\lambda), \exp(i\lambda t) f^{1/2}(\lambda)x)\, d\lambda = \int \exp(-i\lambda t)(f^{1/2}(\lambda)a(\lambda), x)\, d\lambda = 0, \qquad t \in S,$$

and letting

$$b(\lambda) = f^{1/2}(\lambda)a(\lambda),$$

we have the representation

$$a(\lambda) = f^{-1/2}(\lambda)b(\lambda),$$

where $f^{-1/2}(\lambda)$ denotes the inverse operator of the restriction of $f^{1/2}(\lambda)$ on the subspace $X(\lambda) = \overline{f^{1/2}(\lambda)X}$ and the integrable function $b(\lambda)$ is such that

$$b(\lambda) \in f^{1/2}(\lambda)X \quad \text{a.e.,}$$
$$f^{-1/2}(\lambda)b(\lambda) \in L^2(\mathbb{R}^d, X). \tag{1.9}$$

and in addition the Fourier transform $\tilde{b}(t)$, $t \in \mathbb{R}^d$,

$$\tilde{b}(t) = \frac{1}{(2\pi)^d} \int_{\mathbb{R}^d} \exp(-i\lambda t)b(\lambda)\, d\lambda.$$

vanishes on the set S:

$$\tilde{b}(t) = 0, \qquad t \in S. \tag{1.10}$$

We point out here that

$$\|\tilde{b}(t)\| \le \frac{1}{(2\pi)^d} \le \int \|b(\lambda)\|\, d\lambda,$$

$$\int \|b(\lambda)\|\, d\lambda \le \int \|f^{1/2}(\lambda)\| \cdot \|f^{-1/2}(\lambda)b(\lambda)\|\, d\lambda$$

$$\le \left(\int \|f(\lambda)\|\, d\lambda \right)^{1/2} \left(\int \|a(\lambda)\|^2\, d\lambda \right)^{1/2}.$$

Clearly, for any function $b(\lambda)$ of the indicated type, $a(\lambda) = f^{-1/2}(\lambda)b(\lambda)$ belongs to the space $H(\mathbb{R}^d)$—see (1.7)—and $a(\lambda) \perp H(S)$. We arrive at the following representation

$$H(S)^\perp = \overline{f^{-1/2}(\lambda)B(S^c)},$$

where $B(S^c)$ is the collection of all functions $b(\lambda)$ of the form (1.9) with support $\mathrm{Supp}\,\tilde{b} \subseteq S^c$ (cf. (1.10)). We take an open domain $S \subseteq \mathbb{R}^d$ which is bounded or has bounded complement in $\mathbb{R}^d$. For this set

$$H_+(S^c)^\perp = \bigvee_{\varepsilon > 0} H[(S^c)^\varepsilon]^\perp = \bigvee f^{-1/2}(\lambda)B(\overline{S^{-\varepsilon}}),$$

where $S^{-\varepsilon}$ is the complement of $(S^c)^\varepsilon$, an ε-neighborhood of S^c. In other words,

$$H^*(S) = H_+(S^c)^\perp = \overline{f^{-1/2}(\lambda)B(S)}, \tag{1.11}$$

where $B(S)$ is the collection of all functions $b(\lambda)$ of type (1.9) with support

$$\text{Supp } \tilde{b} \subseteq S,$$

since in an open domain S with compact boundary, the closed set $K = \text{Supp } \tilde{b}$ is a positive distance from the boundary ∂S, hence $\text{Supp } \tilde{b} \subseteq S^{-\varepsilon}$ for some $\varepsilon > 0$.

Let a spectral density $f(\lambda)$ in a Hilbert space X be such that there is an associated family of operator-valued, norm-integrable functions of E-exponential type,

$$\varphi_\varepsilon(\lambda) = \int_E \exp(i\lambda t) u(t)\, dt, \qquad \lambda \in \mathbb{R}^d,$$

$$E = \{t: |t_k| \le \varepsilon_k, k = 1, \ldots, d\},$$

(1.12)

and the following properties hold:

$$\int \|f^{-1/2}(\lambda)\varphi_\varepsilon(\lambda)\|^2 \, d\lambda < \infty,$$

(1.13)

the product $f^{-1/2}(\lambda)\varphi_\varepsilon(\lambda)f^{1/2}(\lambda)$ is bounded and

$$\|f^{-1/2}(\lambda)\varphi_\varepsilon(\lambda)f^{1/2}(\lambda) - I\| \to 0$$

(1.14)

as $\varepsilon \to 0$, uniformly in λ in every bounded domain $\subseteq \mathbb{R}^d$. I denotes the identity operator on X.

We consider the system $H^*(S)$, $S \subseteq \mathbb{R}^d$, defined by (1.11) for all open domains S which are bounded or have bounded complements in $\mathbb{R}^d$. We will show that under (1.12)–(1.14) the system will be additive:

$$H^*(S_1 \cup S_2) = H^*(S_1) \vee H^*(S_2).$$

(1.15)

We notice at once that we can always assume that one or the other of the domains S_1, S_2 is bounded. This is because if both the original domains S_1, S_2 are unbounded, then by taking a bounded domain $S_1' = S_1 \cap (S_2^c)^\varepsilon$ we will have $H^*(S_1') \subseteq H^*(S_1)$ and (1.15) will hold if it is true for S_1' and S_2.

Thus, let one of the domains (say, S_1) be bounded. By formula (1.11), functions $a(\lambda) = f^{-1/2}(\lambda)b(\lambda)$ with support $K = \text{Supp } \tilde{b} \subseteq S$ a positive distance from the boundary of the domain $S = S_1 \cup S_2$ form a complete system in the space $H^*(S)$. We take an arbitrary function $a(\lambda) \in H^*(S)$ of this type and look at the corresponding function $b(\lambda)$. It is evident that the set $K = \text{Supp } \tilde{b}$ can be represented as a union of disjoint sets K_1 and $K_2 = K \backslash K_1$ such that

$$\bar{K}_1 \subseteq S_1, \qquad \bar{K}_2 \subseteq S_2$$

and each of them is a positive distance from the boundary of the corresponding domain S_1, S_2. Let

$$\tilde{b}_1(t) = \begin{cases} \tilde{b}(t), & t \in K_1, \\ 0, & t \notin K_1, \end{cases} \qquad \tilde{b}_2(t) = \begin{cases} \tilde{b}(t), & t \in K_2, \\ 0, & t \notin K_2. \end{cases}$$

The bounded (in norm) finite function $\tilde{b}_1(t)$ has a bounded Fourier transform

$$b_1(\lambda) = \int \exp(i\lambda t)\tilde{b}_1(t)\, dt, \qquad \lambda \in \mathbb{R}^d,$$

and for all sufficiently small $\varepsilon > 0$

$$f^{-1/2}(\lambda)\varphi_\varepsilon(\lambda) \in H^*(S_1).$$

In fact, the function $\varphi_\varepsilon(\lambda)b_1(\lambda)$ satisfies condition (1.9) by virtue of the boundedness of $b_1(\lambda)$ and the choice of $\varphi_\varepsilon(\lambda)$—see (1.12), (1.13)—and

$$\frac{1}{(2\pi)^d} \int \exp(-i\lambda t)\varphi_\varepsilon(\lambda)b_1(\lambda)\, d\lambda$$

$$= \sum_k \left(\frac{1}{(2\pi)^d} \int \exp(-i\lambda t)(\varphi_\varepsilon(\lambda)b_1(\lambda), x_k)\, d\lambda \right) x_k,$$

where $\{x_k\}$ is an orthonormal basis of X, and the functions

$$\frac{1}{(2\pi)^d} \int \exp(-i\lambda t)(\varphi_\varepsilon(\lambda)b_1(\lambda), x_k)\, d\lambda$$

$$= \sum_j \frac{1}{(2\pi)^d} \int \exp(-i\lambda t)(b_1(\lambda), x_j)(\varphi_\varepsilon(\lambda)x_j, x_k)\, d\lambda$$

$$= \sum_j \int (u_\varepsilon(t - s)x_j, x_k)(\tilde{b}_1(s), x_j)\, ds, \qquad t \in \mathbb{R}^d,$$

have support in the domain $K^\varepsilon \subseteq S_1$ for small $\varepsilon > 0$ since

$$u_\varepsilon(t) = \frac{1}{(2\pi)^d} \int \exp(-i\lambda t)\varphi_\varepsilon(\lambda)\, d\lambda = 0, \qquad |t| > \varepsilon.$$

Similarly

$$f^{-1/2}(\lambda)\varphi_\varepsilon(\lambda)b(\lambda) \in H^*(S)$$

for small enough $\varepsilon > 0$; we specify here that

$$\|f^{-1/2}(\lambda)\varphi_\varepsilon(\lambda)b(\lambda)\| \le \|f^{-1/2}(\lambda)\varphi_\varepsilon(\lambda)f^{1/2}(\lambda)\| \cdot \|f^{-1/2}(\lambda)b(\lambda)\| \le C\|a(\lambda)\|.$$

As a consequence we get that

$$f^{-1/2}(\lambda)\varphi_\varepsilon(\lambda)b_2(\lambda) = f^{-1/2}(\lambda)\varphi_\varepsilon(\lambda)b(\lambda) - f^{-1/2}(\lambda)\varphi_\varepsilon(\lambda)b_1(\lambda) \in H^*(S_2),$$

because the Fourier transform of the function

$$\varphi_\varepsilon(\lambda)b_2(\lambda) = \varphi_\varepsilon(\lambda)b(\lambda) - \varphi_\varepsilon(\lambda)b_1(\lambda)$$

has support in the domain S_2. We consider the functions

$$a(\lambda) = f^{-1/2}(\lambda)b(\lambda) \quad \text{and} \quad a_\varepsilon(\lambda) = f^{-1/2}(\lambda)\varphi_\varepsilon(\lambda)b(\lambda);$$

their difference is

$$a_\varepsilon(\lambda) - a(\lambda) = (f^{-1/2}(\lambda)\varphi_\varepsilon(\lambda)f^{1/2}(\lambda) - I)a(\lambda)$$

and by (1.14)

$$\int \|a_\varepsilon(\lambda) - a(\lambda)\|^2 \, d\lambda \le \int \|f^{-1/2}(\lambda)\varphi_\varepsilon(\lambda)f^{1/2}(\lambda) - I\|^2 \cdot \|a(\lambda)\|^2 \, d\lambda \to 0$$

as $\varepsilon \to 0$. Thus the function $a(\lambda) = f^{-1/2}(\lambda)b(\lambda)$ in the space $H^*(S)$, $S = S_1 \cup S_2$, belongs to the closed linear span $H^*(S_1) \vee H^*(S_2)$, since the approximating functions $a_\varepsilon(\lambda)$ have a decomposition

$$a_\varepsilon(\lambda) = f^{-1/2}(\lambda)\varphi_\varepsilon(\lambda)b_1(\lambda) + f^{-1/2}(\lambda)\varphi_\varepsilon(\lambda)b_2(\lambda),$$

in which the first and second terms belong to the spaces $H^*(S_1)$ and $H^*(S_2)$ respectively. This proves equation (1.15).

We assume that the space X is finite dimensional and for almost all $\lambda \in \mathbb{R}^d$ the spectral density $f(\lambda)$ is invertible and satisfies the condition

$$\int_0^\infty \frac{\log \sup_{|\lambda| \le r} \|f^{-1}(\lambda)\|}{1 + r^2} \, dr < \infty; \tag{1.16}$$

recall here that for almost all λ, $f(\lambda)$ is a positive trace class operator and when it is invertible, the inverse operator $f^{-1}(\lambda)$ is bounded only when the space X is finite dimensional.

Under condition (1.16) we can show explicitly a family of functions of E-exponential type (1.12), satisfying (1.13)–(1.14).

One can choose a monotonically increasing function $g(r) \ge 1$, with

$$\int_0^\infty \frac{\log g(r)}{1 + r^2} \, dr < \infty,$$

such that for all $\varepsilon > 0$, $r_0 > 0$,

$$\varlimsup_{r \to \infty} \frac{\sup_{|\lambda| \le r} \|f^{-1}(\lambda)\|}{g(\varepsilon r - r_0)} < \infty.$$

The existence of (a majorant) $g(r)$ is clear, for example, if (1.16) is actually satisfied in a somewhat stronger form, namely if †

$$\int_0^\infty \log \sup_{|\mu| \le r} \|f^{-1}(\mu \cdot k(r))\| \, dr < \infty \tag{1.16a}$$

† In connection with the existence of our majorant in the general case see, for example, Ronkin, L. I., *Dokl. Akad. Nauk. USSR*, **5** (1956).

for some monotonically increasing function $k(r)$. For $g(r)$ one can take the function

$$g(r) = \sup_{|\mu| \le r} \| f^{-1}(\mu \cdot k(r)) \|.$$

We temporarily introduce a variable $\lambda \in R^1$ by taking a function $\alpha(z)$ which is analytic in the upper half-plane with boundary values $\alpha(\lambda)$,

$$|\alpha(\lambda)|^2 = \frac{1}{(1 + |\lambda|)^{4d} g(|\lambda|)}, \qquad \lambda \in \mathbb{R}^1,$$

and having Fourier transform

$$\tilde{a}(t) = \frac{1}{2\pi} \int_{-\infty}^{\infty} \exp(-i\lambda t) a(\lambda) \, d\lambda, \qquad t \in \mathbb{R}^1,$$

which vanishes on the negative half-line $t \le 0$; for instance, one can define the function $\alpha(z)$, Im $z > 0$, by the well-known formula

$$\alpha(z) = \exp\left\{ \frac{1}{2\pi i} \int_{-\infty}^{\infty} \log|\alpha(\lambda)|^2 \cdot \frac{1 + \lambda z}{z - \lambda} \frac{d\lambda}{1 + \lambda^2} \right\}.$$

The complex-conjugate function $\bar{\alpha}(z)$ is analytic in the left half-plane Re $z < 0$ with boundary values $\overline{\alpha(\lambda)}$ having Fourier transform $\widetilde{\bar{\alpha}}(t)$ which vanishes on the positive half-line $t \ge 0$. Clearly, the product

$$u(t) = u_0 \exp(i\lambda_0 t)\overline{\alpha(t + 1)}\bar{\alpha}(t - 1), \qquad t \in \mathbb{R}^1,$$

is a bounded function with support

$$\text{Supp } u \subseteq [-1, 1].$$

We choose the parameters u_0 and λ_0 such that

$$\int u(t) \, dt = 1.$$

We have the following expression for the Fourier transform of the function $u(t)$, $t \in \mathbb{R}^1$:

$$\varphi(\lambda) = \int_{|t| \le 1} \exp(i\lambda t)u(t) \, dt$$

$$= u_0 \int_{-\infty}^{\infty} \exp(-i\mu)\alpha(\mu + \lambda_0) \overline{\exp(i(\mu - \lambda))\alpha(\mu - \lambda)} \, d\mu, \qquad \lambda \in \mathbb{R}^1.$$

Keeping the monotonicity of the function $|\alpha(\lambda)|$ in mind, we easily get the following estimates for $|\lambda|/2 \geq |\lambda_0|$:

$$|\varphi(\lambda)| \leq |\mu_0| \int_{-\infty}^{\infty} |\alpha(\mu + \lambda_0)| \cdot |\alpha(\mu - \lambda)| \, d\mu$$

$$\leq |u_0| \cdot \left| \alpha\left(\frac{|\lambda|}{2} - |\lambda_0| \right) \right| \left[\int_{-\infty}^{-|\lambda|/2} |\alpha(\mu - \lambda)| \, d\mu \right.$$

$$\left. + \int_{|\lambda|/2}^{\infty} |\alpha(\mu - \lambda)| \, d\mu \right]$$

$$+ |u_0| \cdot \left| \alpha\left(\frac{|\lambda|}{2} - |\lambda_0| \right) \right| \int_{-|\lambda|/2}^{|\lambda|/2} |\alpha(\mu + \lambda_0)| \, d\mu$$

$$\leq C \cdot \left| \alpha\left(\frac{|\lambda|}{2} - |\lambda_0| \right) \right|$$

$$= \frac{C}{[1 + (|\lambda|/2 - |\lambda_0|)]^{2d}} \cdot \frac{1}{g(|\lambda|/2 - |\lambda_0|)^{1/2}}.$$

Let

$$\varphi_1(\mu) = \varphi(\mu_1) \cdots \varphi(\mu_d), \qquad \mu = (\mu_1, \ldots, \mu_d) \in \mathbb{R}^d.$$

If we take λ equal to the value μ_k for which $|\mu_k| = \max_j |\mu_j|$ among the coordinates of the point $\mu \in \mathbb{R}^d$, then it turns out that for sufficiently large $|\mu|$

$$|\varphi_1(\varepsilon\mu)|^2 \cdot \| f(\mu)^{-1/2} \|^2 \leq C |\varphi(\varepsilon\lambda)|^2 \cdot \| \varphi(\mu)^{-1} \| \leq \frac{C_1}{(1 + \varepsilon|\lambda|/2 - |\lambda_0|)^{4d}},$$

by the initial choice of function $g(r)$, $r > 0$.

It is clear that the functions

$$\varphi_\varepsilon(\lambda) = \varphi_1(\varepsilon\lambda)I, \qquad \lambda \in \mathbb{R}^d, \tag{1.17}$$

satisfy all the conditions (1.12)–(1.14).

Thus, the following proposition holds for a stationary function $\xi(t)$, $t \in \mathbb{R}^d$, and the random field it generates

$$H(S) = \bigvee_{t \in S, \, x \in X} (\xi(t), x), \qquad S \subseteq \mathbb{R}^d, \tag{1.18}$$

under the assumption that for the spectral density $f(\lambda)$, an E-exponential family (1.12) exists (this will be true under condition (1.16)).

Theorem. *The collection*

$$H^*(S) = H_+(S^c)^\perp, \qquad S \subseteq T,$$

defined for all open domains $S \subseteq \mathbb{R}^d$ which are bounded or have bounded complement $S^c = \mathbb{R}^d \backslash S$, is additive.

Earlier we called such a collection the dual random field for $H(S), S \subseteq T$ — see Chapter 3, §3.

§2. The Markov Property for Stationary Functions

1. The Markov Property When a Dual Field Exists

In considering the general Markov property of a random field

$$H(S), \quad S \subseteq T = \mathbb{R}^d,$$

in Chapter 2, §3, we were concerned with complementary domains

$$S_1 = S, \qquad S_2 = \mathbb{R}^d \backslash \bar{S},$$

with separating boundary $\Gamma = \partial S$ for which $H(\Gamma^\varepsilon)$, $\varepsilon > 0$, split the spaces $H(S_1)$ and $H(S_2)$.

Assuming that one of the domains S_1, S_2 is bounded, we will construct a separating boundary Γ in the following way. Let L be some (bounded) neighborhood of the point $t = 0$ and let $S + L$ denote the set of all points of the form $s + t$, $s \in S$, $t \in L$; Let

$$\Gamma = \overline{(S_1 + L)} \cap \overline{(S_2 + L)}. \tag{2.1}$$

We will call a random field *L-Markov* if it is Markov with respect to all domains $S \subseteq \mathbb{R}^d$ which are bounded or have bounded complement and have separating boundary of the form (2.1).

We note that a random field is Markov under our basic definition if it is *L*-Markov for any neighborhood L; formally, if $L = \{0\}$.

By results in Chapter 2, §3, the *L*-Markov property is equivalent to a related orthogonality property of the dual field, namely,

$$H^*(S_1^{-\varepsilon}) \perp H^*(S_2^{-\varepsilon}), \qquad \varepsilon > 0,$$

where we have introduced, for brevity, the notation

$$S_1^{-\varepsilon} = (\bar{S}_2 \cup \Gamma^\varepsilon)^c, \qquad S_2^{-\varepsilon} = (\bar{S}_1 \cup \Gamma^\varepsilon)^c.$$

For a stationary function $\xi(t)$, $t \in \mathbb{R}^d$, with spectral density $f(\lambda)$ on a Hilbert space X, this orthogonality property can be expressed by the equation

$$\int (f^{-1/2}(\lambda)b_1(\lambda), f^{-1/2}(\lambda)b_2(\lambda)) \, d\lambda = 0, \qquad b_1 \in B(S_1^{-\varepsilon}), b_2 \in B(S_2^{-\varepsilon}), \tag{2.2}$$

cf. (1.11).

The term *generalized Fourier transform* will denote the operator $\widetilde{f^{-1}}(t)$ giving the relation

$$\int (f^{-1/2}(\lambda)b_1(\lambda), f^{-1/2}(\lambda)b_2(\lambda))\, d\lambda = \iint (\widetilde{f^{-1}}(s-t)\tilde{b}_1(s), \tilde{b}_2(t))\, ds\, dt$$

for all pairs of functions $\tilde{b}_1(s), \tilde{b}_2(t)$ which are Fourier transforms of functions $b_1(\lambda), b_2(\lambda)$ of type (1.9), for which at least one of the supports

$$K_1 = \operatorname{Supp} \tilde{b}_1, \qquad K_2 = \operatorname{Supp} \tilde{b}_2$$

is compact. Formally, our "generalized function" $\widetilde{f^{-1}}$ is defined for pairs of variables (s, t)—or more accurately, for pairs of "basic functions" $(\tilde{b}_1, \tilde{b}_2)$. However, it is not changed by the shift $(s, t) \to (s + u, t + u), u \in \mathbb{R}^d$, and this allows us to regard it as depending on the difference $s - t$.

Let S be some domain in $\mathbb{R}^d$. We will consider that $\widetilde{f^{-1}}(t_1 - t_2) = 0$ for $t_1 - t_2 \in S$ if $\widetilde{f^{-1}}$ vanishes for all "basic functions" $(\tilde{b}_1, \tilde{b}_2)$ with supports $\operatorname{Supp} \tilde{b}_1 = K_1, \operatorname{Supp} \tilde{b}_2 = K_2$ such that

$$\overline{K_1 - K_2} = \overline{K_1 + (-K_2)} \subseteq S.$$

Condition (2.2) is equivalent to $\widetilde{f^{-1}}(t) = 0$ outside the closed set $\overline{L - L}$, in other words, the support of our generalized function $\widetilde{f^{-1}}$ lies in this domain:

$$\operatorname{Supp} \widetilde{f^{-1}} \subseteq \overline{L - L}. \tag{2.3}$$

To see this, let condition (2.2) hold. We take an arbitrary pair (b_1, b_2) with supports $K_1 = \operatorname{Supp} \tilde{b}_1, K_2 = \operatorname{Supp} \tilde{b}_2$, such that

$$\overline{K_1 - K_2} \subseteq \mathbb{R}^d \backslash (\overline{L - L}).$$

By choosing $\delta > 0$ such that 2δ is smaller than the distance from the closed set $\overline{K_1 - K_2}$ to the compact set $(\overline{L - L})$ and taking $\varepsilon < \delta$, we get that K_1 and K_2 lie outside the closures of the domains $S_2 \cup \Gamma^\varepsilon \subseteq (S_2 + L)^\varepsilon$ and $S_1 \cup \Gamma^\varepsilon \subseteq (S_1 + L)^\varepsilon$ when $S_1 = (K_1 - L)^\delta$, i.e.,

$$K_1 \subseteq S_1^{-\varepsilon}, \qquad K_2 \subseteq S_2^{-\varepsilon}.$$

This is clear from the inequality

$$|s_2 - (s_1 - t_1 + t_2)| = |(s_1 - s_2) - (t_1 - t_2)| \geq 2\delta > \delta + \varepsilon$$

for all $s_1 \in K_1, s_2 \in K_2$ and $t_1, t_2 \in L$, and the inequality

$$|s_1 - (t_1 + t_2)| = |t_1 - (s_1 - t_2)| \geq \delta$$

for all $s_1 \in K_1, t_1 \in S_2$ and $t_2 \in L$, (recall that S_2 is the complement of $\bar{S}_1$). Thus the inclusion (2.3) is true under condition (2.2). Conversely, for any complementary domains S_1, S_2 and functions (b_1, b_2) with

$$K_1 = \operatorname{Supp} \tilde{b}_1 \subseteq S_1^{-\varepsilon}, \qquad S_1^{-\varepsilon} = (\overline{S_2 \cup \Gamma^\varepsilon})^c,$$

$$K_2 = \operatorname{Supp} \tilde{b}_2 \subseteq S_2^{-\varepsilon}, \qquad S_2^{-\varepsilon} = (\overline{S_1 \cup \Gamma^\varepsilon})^c,$$

where

$$\Gamma = (\overline{S_1 + L}) \cap (\overline{S_2 + L}),$$

we have

$$\overline{S_1^{-\varepsilon}} - L = \overline{S_1^{-\varepsilon}} - \overline{L} \subseteq S_1, \qquad \overline{S_2^{-\varepsilon}} - L = \overline{S_2^{-\varepsilon}} - \overline{L} \subseteq S_2;$$

for instance, if $s_1 \in \overline{S_1^{-\varepsilon}}$ then $s_1 \notin \Gamma$ and $s_1 - \overline{L}$ lies in S_1 because otherwise $s_1 - t \in \overline{S_2}$ for some $t \in \overline{L}$ and

$$s_1 = (s_1 - t) + t \in (\overline{S_2 + L}) \cap (\overline{S_1 + L}) = \Gamma.$$

As a result the closed sets

$$\overline{K_1 - L} = K_1 - \overline{L} \subseteq S_1, \qquad \overline{K_2 - L} = K_2 - \overline{L} \subseteq S_2$$

do not intersect and since

$$|(s_1 - t_1) - (s_2 - t_2)| = |(s_1 - s_2) - (t_1 - t_2)| \geq \rho > 0$$

for all $s_1 \in K_1$, $s_2 \in K_2$, and $t_1, t_2 \in \overline{L}$, we see that $\overline{K_1 - K_2} = K_1 - K_2$ is disjoint from $\overline{L - L} = \overline{L} - \overline{L}$ (recall here that the set $\overline{L}$ and one of $K_1 \subseteq \overline{S_1^{-\varepsilon}}$, $K_2 \subseteq \overline{S_2^{-\varepsilon}}$ are compact). Thus the supports K_1, K_2 of the functions $\tilde{b}_1$, $\tilde{b}_2$ are such that a generalized function of the type (2.3) vanishes on $(\tilde{b}_1, \tilde{b}_2)$ and equation (2.2) holds.

Thus, for a stationary function $\xi(t)$, $t \in \mathbb{R}^d$, and the stationary field (1.18) it generates, the following proposition is true when the conjugate field is available.

Theorem 1. *Condition (2.3) is necessary and sufficient for the stationary field with spectral density $f(\lambda)$ to be L-Markov.*

As a corollary we have the stationary field Markov if and only if

$$\text{Supp } f^{-1} = \{0\}. \tag{2.4}$$

2. Analytic Markov Conditions

For a spectral density $f(\lambda)$, let there exist an E-exponential family of functions

$$\varphi_\varepsilon(\lambda) = \int_E \exp(i\lambda t) u_2(t) \, dt, \qquad E = \{t : |t_k| < \varepsilon_k, k = 1, \ldots, d\},$$

with properties (1.12)–(1.14).

Lemma. *A stationary field is L-Markov if and only if the integrable function*

$$g_\varepsilon(\lambda) = \varphi_\varepsilon^*(\lambda) f^{-1}(\lambda) \varphi_\varepsilon(\lambda)$$

has a Fourier transform with support

$$\text{Supp } g_\varepsilon(t) \subseteq \overline{(L + E) - (L + E)}. \tag{2.5}$$

PROOF. Let a stationary field be L-Markov. Then (2.3) holds. We take arbitrary $x, y \in X$ and functions

$$b_1(\lambda) = \exp(i\lambda t)\varphi_\varepsilon(\lambda)x \in B(K_1), \qquad K_1 = E + t,$$
$$b_2(\lambda) = \varphi_\varepsilon(\lambda)y \in B(K_2), \qquad K_2 = E.$$

We have

$$\int \exp(i\lambda t)(g_\varepsilon(\lambda)x, y) \, d\lambda = \int (f^{-1/2}(\lambda)b_1(\lambda), f^{-1/2}(\lambda)b_2(\lambda)) \, d\lambda = 0,$$

$$t \notin \overline{(L + E) - (L + E)},$$

because for any t which is a distance $\delta > 0$ from the set $(L + E) - (L + E)$,

$$|t - [(s_2 - s_1) - (t_1 - t_2)]| = |[(t + s_1) - s_2] - (t_2 - t_1)| \geq \delta > 0$$

for all $s_1, s_2 \in E$ and $t_1, t_2 \in L$, hence the set $\overline{K_1 - K_2}$ lies outside $\overline{L - L}$.

Deriving condition (2.3) from (2.5) is somewhat more complex. For a function with bounded $K = \text{Supp } \tilde{b}(t)$ the integrable function $\|b(\lambda)\|$ is bounded and square-integrable:

$$\frac{1}{(2\pi)^d} \int \|b(\lambda)\|^2 \, d\lambda = \sum_k \frac{1}{(2\pi)^d} \int |(b(\lambda), x_k)|^2 \, d\lambda = \sum_k \int |(b(t), x_k)|^2 \, dt$$

$$= \int \|\tilde{b}(t)\|^2 \, dt,$$

where $\{x_k\}$ is an orthonormal basis of X; in addition

$$\|\tilde{b}(t)\| \leq \frac{1}{(2\pi)^d} \int \|b(\lambda)\| \, d\lambda < \infty,$$

$$\|b(\lambda)\| = \left(\sum_k |(b(\lambda), x_k)|^2 \right)^{1/2} \leq \int \|\tilde{b}(t)\| \, dt < \infty.$$

For almost all fixed $\lambda \in \mathbb{R}^d$ the operator $f^{-1/2}(\lambda)\varphi_\varepsilon(\lambda)$ is bounded on X and we have

$$b(\lambda) = \sum_k (b(\lambda), x_k)x_k,$$

$$f^{-1/2}(\lambda)\varphi_\varepsilon(\lambda)b(\lambda) = \sum_k (b(\lambda), x_k)f^{-1/2}(\lambda)\varphi_\varepsilon(\lambda)x_k;$$

moreover here we not only have convergence almost everywhere in X but also in the space $L^2(\mathbb{R}^d, X)$ since by (1.13) the function

$$\|g_\varepsilon(\lambda)\| = \|f^{-1/2}(\lambda)\varphi_\varepsilon(\lambda)\|^2$$

is integrable and the partial sums $\sum_{k=m}^n |(b(\lambda), x_k)|^2$ are uniformly bounded, so that

$$\int \left\| \sum_{k=m}^n (b(\lambda), x_k)f^{-1/2}(\lambda)\varphi_\varepsilon(\lambda)x_k \right\|^2 \, d\lambda \leq \int \|g_\varepsilon(\lambda)\| \sum_{k=m}^n |(b(\lambda), x_k)|^2 \, d\lambda \to 0$$

as $m, n \to \infty$. Now take arbitrary

$$b(\lambda) = b_1(\lambda) \in B(K_1), \qquad b_2(\lambda) \in B(K_2),$$

with supports such that the closed set $\overline{K_1 - K_2}$ lies outside the compact set $L - L$. We have

$$\int (f^{-1/2}(\lambda)b_1(\lambda), f^{-1/2}(\lambda)b_2(\lambda)) \, d\lambda$$

$$= \lim_{\varepsilon \to 0} \int (f^{-1/2}(\lambda)\varphi_\varepsilon(\lambda)b_1(\lambda), f^{-1/2}(\lambda)\varphi_\varepsilon(\lambda)b_2(\lambda)) \, d\lambda$$

where

$$f^{-1/2}(\lambda)\varphi_\varepsilon(\lambda)b_1(\lambda) = \sum_k (b_1(\lambda), x_k) f^{-1/2}(\lambda)\varphi_\varepsilon(\lambda)x_k,$$

and to show that (2.3) follows from (2.5) it is sufficient to show that

$$\int (b_1(\lambda), x_k)(f^{-1/2}(\lambda)\varphi_\varepsilon(\lambda)x_k, f^{-1/2}(\lambda)\varphi_\varepsilon(\lambda)b_2(\lambda)) \, d\lambda$$

$$= \int (b_1(\lambda), x_k)(g_\varepsilon(\lambda)x_k, b_2(\lambda)) \, d\lambda = 0$$

for small $\varepsilon > 0$. We have

$$|(b_1(\lambda), x_k)| \leq \|b_1(\lambda)\| \leq C,$$

$$\sum_j |(g_\varepsilon(\lambda)x_k, x_j)\overline{(b_2(\lambda), x_j)}| \leq \|g_\varepsilon(\lambda)\| \, \|f^{-1/2}(\lambda)b_2(\lambda)\| \, \|f(\lambda)\|_1^{1/2};$$

the last two factors are square-integrable and by (2.5) the norm $\|g_\varepsilon(\lambda)\|$ is bounded since the support $\operatorname{Supp} \tilde{g}_\varepsilon(t)$ is bounded. Thus

$$\int (b_1(\lambda), x_k)(g_\varepsilon(\lambda)x_k, b_2(\lambda)) \, d\lambda$$

$$= \sum_j \int (b_1(\lambda), x_k)(g_\varepsilon(\lambda)x_k, x_j)\overline{(b_2(\lambda), x_j)} \, d\lambda$$

$$= \sum_j \iint (\tilde{b}_1(s), x_k)(\tilde{g}_\varepsilon(s - t)x_k, x_j)(\tilde{b}_2(t), x_j) \, ds \, dt = 0$$

for all $\varepsilon > 0$, for which the closed set $\overline{K_1 - K_2}$ lies outside the compact set $\overline{(L + E) - (L + E)}$. This proves the lemma. $\qquad\square$

Let us look at the L-Markov property in the case of a neighborhood of the type

$$L = \{t \colon |t_k| \leq L_k, k = 1, \dots, d\}.$$

Theorem 2. *A stationary random function with spectral density $f(\lambda)$ is L-Markov if $f^{-1}(\lambda)$ is an $(L - L)$-exponential entire function.*

Recall that for a neighborhood L of the type above, an entire analytic function $g(z)$ of several complex variables $z = (z_1, \ldots, z_d)$ is called *L-exponential* if for any $\delta > 0$ the bound

$$\|g(z)\| \leq C_\delta \exp\left\{\sum_{k=1}^d (L_k + \delta)|z_k|\right\}$$

is true for appropriate constant C_δ. We make use of the well-known fact that if an L-exponential function is integrable as a function $g(\lambda)$, $\lambda \in \mathbb{R}^d$, of real variables $\lambda = (\lambda_1, \ldots, \lambda_d)$ then it has a Fourier transform with support

$$\widetilde{g(t)} \subseteq \{t : |t_k| \leq L_k, k = 1, \ldots, d\}.$$

It is clear that in the case of an $(L - L)$-exponential function $f^{-1}(\lambda)$, where

$$L - L = \{t : |t_k| < 2L_k, k = 1, \ldots, d\},$$

the corresponding

$$g_\varepsilon(\lambda) = \varphi_\varepsilon^*(\lambda) f^{-1}(\lambda)\varphi_\varepsilon(\lambda)$$

are $[(L + E) - (L + E)]$-exponential functions associated with neighborhoods

$$(L + E) - (L + E) = \{t : |t_k| < 2(L_k + \varepsilon_k), k = 1, \ldots, d\};$$

Hence condition (2.5) of our lemma on the L-Markov property is satisfied.

Formally letting $L = \{0\}$, we get the following proposition from Theorem 2:

A stationary random function with spectral density $f(\lambda)$ is Markov if $f^{-1}(\lambda)$ is an entire function of minimal exponential type.

We should keep in mind here that we are referring to the Markov property with respect to all domains $S \subseteq \mathbb{R}^d$ which are bounded or have bounded complements, for which we established earlier a condition for duality—see the theorem in §1.2.

We will take a more detailed look at the case of a spectral density $f(\lambda)$ of the form

$$\int \frac{\|f^{-1}(\lambda)\|}{(1 + |\lambda|^2)^\ell} d\lambda < \infty. \tag{2.6}$$

In this case the generalized Fourier transform $\widetilde{f^{-1}}(t)$ can be realized as the generalized function

$$(u, \widetilde{f^{-1}}) = \frac{1}{(2\pi)^d} \int \tilde{u}(\lambda) f^{-1}(\lambda) \, d\lambda, \qquad u \in C_0^\infty(\mathbb{R}^d),$$

where

$$\tilde{u}(\lambda) = \int \exp(i\lambda t)u(t)\, dt.$$

We remark that the generalized operator-valued function $\widetilde{f^{-1}}$ can be defined with the help of the scalar generalized functions

$$\widetilde{f^{-1}_{x,y}} = (\widetilde{f^{-1}}x, y), \qquad x, y \in X,$$

each of which is a standard (generalized) Fourier transform:

$$(u, \widetilde{f^{-1}_{x,y}}) = \frac{1}{(2\pi)^d} \int \tilde{u}(\lambda)(f^{-1}x, y)\, d\lambda, \qquad u \in C_0^\infty(\mathbb{R}^d).$$

We will show that condition (2.3) derived earlier remains in force if $\widetilde{f^{-1}}$ is interpreted this way; we will call it a standard Fourier transform.

In fact, taking vector-valued functions $b_1(\lambda) = \tilde{u}_1(\lambda)x$ and $b_2(\lambda) = \tilde{u}_2(\lambda)y$ for scalar $u_1, u_2 \in C_0^\infty(\mathbb{R}^d)$ and arbitrary $x, y \in X$, we get from (2.3)

$$\int (f^{-1/2}(\lambda)b_1(\lambda), f^{-1/2}(\lambda)b_2(\lambda))\, d\lambda = \int \tilde{u}_1(\lambda)\overline{\tilde{u}_2(\lambda)}(f^{-1}(\lambda)x, y)\, d\lambda$$

$$= (u_1 * u_2, \widetilde{f^{-1}_{x,y}}) = 0$$

for supports $K_1 = \mathrm{Supp}\, u_1$, $K_2 = \mathrm{Supp}\, u_2$ such that $K_1 - K_2$ lies outside $\overline{L - L}$. As a result, for a standard Fourier transform $\widetilde{f^{-1}}$ we have the inclusion

$$\mathrm{Supp}\, \widetilde{f^{-1}}(t) \subseteq \overline{L - L}. \tag{2.7}$$

On the other hand, from this new condition we get back condition (2.3) with its former meaning. One can prove this, for example, with the help of our lemma on the L-Markov property, in which we replace the spectral density $f(\lambda)$ of type (2.6) (by an E-exponential family of the form

$$\varphi_\varepsilon(\lambda) = \tilde{u}(\varepsilon\lambda)I,$$

for any function $u \in C_0^\infty(\mathbb{R}^d)$ with support in the domain $|t| \leq 1$ and $\int u(t)\, dt = 1$. It is then evident that (2.7) implies (2.5), which is equivalent to our earlier condition (2.3).

This is also true for the Markov property (2.4). It is clear that one can express this Markov condition in the form

$$\mathrm{Supp}\, \widetilde{f^{-1}_{x,y}} = \{0\}, \qquad x, y \in X, \tag{2.8}$$

and, as is well known, condition (2.8) implies that for a function $f^{-1}(\lambda)$ of type (2.6), $(f^{-1}(\lambda)x, y)$, $x, y \in X$, is a polynomial in $\lambda \in \mathbb{R}^d$.

We formulate this as a proposition:

Theorem 3. *A stationary function $\xi(t)$, $t \in \mathbb{R}^d$, with values in the Hilbert space X and with spectral density $f(\lambda)$ of the type (2.6) is Markov if and only if for any $x, y \in X$, the scalar product $(f^{-1}(\lambda)x, y)$ is a polynomial in $\lambda \in \mathbb{R}^d$.*

§3. Markov Extensions of Random Processes

1. Minimal Nonanticipating Extension

By a random process we mean a space-valued function

$$H(t) = U_t X, \qquad t \in T, \tag{3.1}$$

of a real parameter t, thought of as time, running over a set $T \subseteq \mathbb{R}^1$; here U_t denotes a family of continuous linear operators from some Hilbert space X into the space $L^2(\Omega, \mathscr{A}, P)$ of random variables on the probability space $(\Omega, \mathscr{A}, P)$.

We introduce the *covariance function*

$$B(t, s) = U_t^* U_s, \qquad s, t \in T; \tag{3.2}$$

note that

$$(B(t, s)x, y) = (U_s x, U_t y), \qquad x, y \in X.$$

In speaking of a Markov process (3.1) we will mean the usual Markov property (in the wide sense), expressed as

$$P(H_{-\infty, t}) H_{t, \infty} = H(t), \qquad t \in T; \tag{3.3}$$

for convenience we have altered our notation slightly:

$$H_{-\infty, t} = H(-\infty, t) \qquad H_{t, \infty} = H(t, \infty), \qquad P(H_{-\infty, t}) = P(-\infty, t).$$

where $P(H)$ denotes the projection on the corresponding space H, cf., Chapter 2 (1.14).

We assume that for all t the operator $B(t, t) = U_t^* U_t$ is invertible. Letting

$$\tilde{U}_t = U_t B^{-1/2}(t, t), \qquad \tilde{H}(t) = \tilde{U}_t X,$$

we would be dealing with the spaces $\tilde{H}(t) = H(t)$, $t \in T$, and the covariance function

$$\tilde{B}(t, t) = B^{-1/2}(t, t) B(t, t) B^{-1/2}(t, t) = I.$$

Thus in considering a random process (3.1) we will assume that,

$$B(t, t) = I, \qquad t \in T.$$

We will show that the Markov property (3.3) is equivalent to the following semi-group property of the covariance function:

$$B(s, u) = B(s, t) B(t, u), \qquad s \leq t \leq u. \tag{3.4}$$

In fact, the condition $B(t, t) = U_t^* U_t = I$ implies that the operator U_t is an isometry and its adjoint operator U_t^* is

$$U_t^* h = \begin{cases} U_t^{-1} h, & h \in H(t), \\ 0, & h \in H(t)^\perp, \end{cases}$$

so that $U_t U_t^* = P_t$ is the projection on $H(t)$; we can express the Markov property (2.3) by the equation

$$(U_u x - P_t U_u x, U_s y) = 0, \qquad s \le t \le u, x, y \in X,$$

or what is in effect equation (3.4):

$$U_s^* U_u - U_s^*(U_t U_t^*)U_u = 0, \qquad s \le t \le u.$$

We have the following result:

Theorem 1. *A random process is Markov if and only if its covariance function satisfies equation* (3.4).

We will call a random process $\tilde{H}(t)$, $t \in T$, an *extension* of the process $H(t)$, $t \in T$, if we have the inclusion

$$H(t) \subseteq \tilde{H}(t), \qquad t \in T.$$

The question naturally arises of the existence of a minimal Markov extension $H^0(t)$, $t \in T$, having the property that

$$H^0(t) \subseteq \tilde{H}(t), \qquad t \in T,$$

for any Markov extension in some class that we are considering. We note that this question can not be posed in connection with all Markov extensions. For instance, for a triple H_1, H_2, H_3 a natural minimal extension in the general case would be

$$H_1^0 = H_1, \qquad H_2^0 = H_1 \vee H_2, \qquad H_3^0 = H_3,$$

but what minimality property would it have in relation to the Markov extension

$$\tilde{H}_1 = H_1, \qquad \tilde{H}_2 = H_2 \vee H_3, \qquad \tilde{H}_3 = H_3?$$

We will call an extension $H(t)$, $t \in T$, *nonanticipating* if

$$\overline{P(\tilde{H}_{-\infty,t})H_{t,\infty}} = \overline{P(H_{-\infty,t})H_{t,\infty}}. \tag{3.5}$$

A trivial example of a nonanticipating Markov extension is $\tilde{H}(t) = H_{-\infty,t}$, $t \in T$.

Theorem 2. *For any process $H(t)$, $t \in T$, there is a minimal nonanticipating Markov extension $H^0(t)$, $t \in T$; this extension is formed by the spaces*

$$H^0(t) = \overline{P(H_{-\infty,t})H_{t,\infty}}. \tag{3.6}$$

PROOF. The minimality of the extension (3.6) is clear since for any non-anticipating Markov extension $\tilde{H}(t)$, $t \in T$, we have

$$\tilde{H}(t) = \overline{P(\tilde{H}_{-\infty,t})\tilde{H}_{t,\infty}} \supseteq \overline{P(\tilde{H}_{-\infty,t})H_{t,\infty}} = \overline{P(H_{-\infty,t})H_{t,\infty}} = H^0(t).$$

We will show that the random process $H^0(t)$, $t \in T$, is Markov. By definition $H^0_{t,\infty}$ is the closed linear span of the spaces $H^0(u) = \overline{P(H_{-\infty,u})H_{u,\infty}}$, $u \geq t$. Taking any nonanticipating Markov extension $H(t)$, $t \in T$, and using (3.3) we get

$$P(H^0_{-\infty,t})H^0(u) = P(H^0_{-\infty,t})\overline{P(\tilde{H}_{-\infty,u})H_{u,\infty}}$$

$$\subseteq P(H^0_{-\infty,t})\overline{P(\tilde{H}_{-\infty,t})\overline{P(\tilde{H}_{-\infty,u})H_{u,\infty}}}$$

$$\subseteq P(H^0_{-\infty,t})\overline{P(\tilde{H}_{-\infty,t})H_{u,\infty}} \subseteq P(H^0_{-\infty,t})\overline{P(\tilde{H}_{-\infty,t})H_{t,\infty}}$$

$$\subseteq P(H^0_{-\infty,t})H^0(t) = H^0(t).$$

Thus

$$P(H^0_{-\infty,t})H^0_{t,\infty} \subseteq H^0(t),$$

which shows that the Markov property holds for the process $H^0(t)$, $t \in T$.

$\square$

2. Markov Stationary Processes

We turn to the general stationary (in a broad sense) process $H(t)$ on the real line $T = (-\infty, \infty)$; the stationarity implies the presence of a continuous group of unitary operators U_t which "shift" the spaces $H(t)$ in a time flow:

$$H(t) = U_t X, \qquad X = H(0), \qquad -\infty < t < \infty, \tag{3.7}$$

(we are assuming that the operators U_t are defined on the closed linear span of the spaces $H(t)$, $-\infty < t < \infty$).

If we let $X = H(0)$ play the role of a "parameter" space, we can characterize the stationary process being considered with the help of the covariance function $B(t)$, $-\infty < t < \infty$, whose value at each moment of time t is an operator $B(t)$ on the space X of the form

$$B(t) = PU_t P,$$

where P is projection on X. Looking at the restriction of the operators U_t, $-\infty < t < \infty$, on space $X = H(0)$, it is clear that as in (3.2) we have

$$B(t, s) = B(s - t) = U_t^* U_s,$$

and for the coupled scalar components

$$\xi_x(t) = U_t x, \qquad -\infty < t < \infty, x \in X,$$

of the random process $H(t) = U_t X$ we have

$$(B(t - s)x, y) = (\xi_x(t), \xi_y(s)), \qquad -\infty < s, t < \infty. \tag{3.8}$$

We can represent the covariance function $B(t)$ as a Fourier integral of some positive operator-valued measure $dF(\lambda)$:

$$B(t) = \int_{-\infty}^{\infty} \exp(i\lambda t)F(d\lambda). \tag{3.9}$$

This measure, called the *spectral measure* of the stationary process, is closely related to the spectral family $E(d\lambda)$, $-\infty < \lambda < \infty$, in the Stone representation of the unitary group U_t,

$$U_t = \int_{-\infty}^{\infty} \exp(i\lambda t)E(d\lambda), \qquad -\infty < t < \infty;$$

namely,

$$F(d\lambda) = PE(d\lambda)P.$$

Let $H(t)$, $-\infty < t < \infty$, be a stationary Markov process. Its covariance function has the semi-group property:

$$B(t + s) = B(t)B(s), \qquad s, t \geq 0, \tag{3.10}$$

cf. (3.4); moreover,

$$B(-t) = B(t)^*.$$

We suppose that the covariance function $B(t)$, $t \geq 0$, is continuous in operator norm. Then, being a solution of (3.10), it is known that it can be represented in the form

$$B(t) = \exp(-at), \qquad t \geq 0, \tag{3.11}$$

where a is some linear operator; in particular, (3.11) holds in case the space X is finite dimensional.

Recall how one proves formula (3.11). From the continuity

$$\|B(t) - I\| \to 0, \qquad t \to 0,$$

there exists a bounded inverse operator $B(t)^{-1}$ for sufficiently small t. For sufficiently close t_1, t_2 the operator $\int_{t_1}^{t_2} B(s)\, ds$ has a bounded inverse for the same reason:

$$\left\| \frac{1}{t_2 - t_1} \int_{t_1}^{t_2} B(s)\, ds - B(t) \right\| \to 0 \quad \text{as } t_1, t_2 \to t.$$

From equation (3.10),

$$\frac{B(h) - I}{h} = \left[\frac{1}{h} \int_{t_2}^{t_2 + h} B(s)\, ds - \frac{1}{h} \int_{t_1}^{t_1 + h} B(s)\, ds \right] \cdot \left[\int_{t_1}^{t_2} B(s)\, ds \right]^{-1}.$$

From the expression on the right side it is evident that the following limit exists (in operator norm)

$$\lim_{h \to 0} \frac{B(h) - I}{h} = -a;$$

the limit thus defined is the so-called *infinitesimal generator* for the semi-group $B(t)$, $t \geq 0$. With it, equation (3.10) leads to the differential equation

$$\frac{d}{dt} B(t) = -aB(t), \qquad t > 0,$$

$$B(0) = I,$$

whose solution is the indicated function (3.11).

Notice that our covariance function $B(t)$ satisfies the condition $\|B(t)\| \leq 1$. From (3.10) it follows that if the norms $\|B(t)\|$ are not identically 1 then the function $\|B(t)\|$, $t \to \infty$, decays exponentially and there is a Fourier transform of the function

$$B(t) = \begin{cases} \exp(-at), & t \geq 0, \\ \exp(-a^*t), & t \leq 0, \end{cases}$$

which gives us a *spectral density* $f(\lambda) = F(d\lambda)/d\lambda$:

$$f(\lambda) = \frac{1}{2\pi} \int_{-\infty}^{\infty} \exp(-i\lambda t) B(t) \, dt$$

$$= \frac{1}{2\pi} \left[\int_{-\infty}^{0} \exp(i\lambda I - a^*)t \, dt + \int_{0}^{\infty} \exp(i\lambda I - a)t \, dt \right]$$

$$= \frac{1}{2\pi} \cdot (i\lambda I + a)^{-1}(a + a^*)(i\lambda I + a^*)^{-1}; \tag{3.12}$$

the inverse operators $(i\lambda I + a)^{-1}$ and $(-i\lambda I + a^*)^{-1}$ exist because the operator a must have pure imaginary eigenvalues, just from the condition $\|\exp(-at)\| \equiv 1$.

In particular, formula (3.12) shows that in the finite dimensional case, where it is a matter of a matrix-valued spectral density $f(\lambda)$, the elements are rational functions.

Let us look at a finite dimensional stationary process $H(t)$, $-\infty < t < \infty$, (each of the spaces $H(t)$ is finite dimensional). We will assume that it is regular:

$$\bigcap_t H_{-\infty, t} = 0.$$

Then its minimal Markov extension $H^0(t)$, $-\infty < t < \infty$, is also a regular stationary process by (3.6).

Suppose that the process $H^0(t)$, $-\infty < t < \infty$, is finite dimensional. Then its covariance function $B^0(t)$ will have the property that

$$\|B^0(t)\| \to 0, \qquad t \to \infty,$$

because by the regularity condition, for any fixed $x, y \in X^0 = H^0(0)$ we have $(U_t x, y) \to 0$, $t \to -\infty$. Consequently the process $H^0(t)$, $-\infty < t < \infty$, has a (matrix-valued) spectral density $f^0(\lambda)$ of the form (3.12) with rational elements. It is clear that

$$B(t) = PB^0(t)P,$$

where P is the projection operator on the space $H(0)$ and

$$f(\lambda) = Pf^0(\lambda)P$$

is the spectral density for the original process $H(t)$, $-\infty < t < \infty$. Furthermore, in the matrix representation of $f(\lambda)$ all the elements will be rational functions. In speaking of a matrix-valued spectral density $f(\lambda)$, we should keep in mind that we have chosen a basis $x_1, \ldots, x_n$ of the "parameter" space X and that the operator-valued function is a matrix with elements

$$f_{kj}(\lambda) = (f(\lambda)x_k, x_j), \qquad j, k = 1, \ldots, n;$$

when all the elements are rational functions of λ, $-\infty < \lambda < \infty$, we will call the matrix-valued function $f(\lambda)$ *rational*.

As is well known, each finite dimensional stationary process having a rational spectral density $f(\lambda)$ admits a finite-dimensional nonanticipating Markov extension.

We will dwell briefly on one of the extensions which arises from a suitable factorization of the integrable rational matrix-valued function $f(\lambda)$ of rank m:

$$f(\lambda) = \frac{1}{2\pi} P(i\lambda)^* \frac{1}{|L(i\lambda)|^2} P(i\lambda),$$

where $P(z)$ is a polynomial of degree $\leq l - 1$ with $(n \times m)$-matrix coefficients, and $L(z)$ is a scalar polynomial of degree l, all of whose zeros lie in the left half-plane. (This factorization is connected with a "renewal process" in a way which is well understood.) The extension is the following. One chooses m "white nose" process

$$\dot{\eta}_j(t), \qquad j = 1, \ldots, m,$$

which are mutually orthogonal and which generate the "past" of our stationary process with components

$$\xi_k(t) = U_t x_k, \qquad -\infty < t < \infty, k = 1, \ldots, n;$$

more precisely, for each t the corresponding space $H_{-\infty, t}$ is the closed linear span of the variables

$$(u, \dot{\eta}_j), \qquad u \in C_0^\infty(-\infty, t), \qquad j = 1, \ldots, m.$$

Furthermore, one defines stationary functions

$$\tilde{\xi}_j(t) = U_t \tilde{\xi}_j(0), \qquad j = 1, \ldots, m,$$

each of which is connected to a "white noise" $\dot{\eta}(t) = \dot{\eta}_j(t)$ stochastic differential equation—Chapter 2, (1.21)—associated with the operator $L = L(d/dt)$. The stationary $(l \times m)$-dimensional random process $\tilde{H}(t)$, $-\infty < t < \infty$, with $\tilde{H}(t)$ the closure of the values of the random functions $\tilde{\xi}_j(t)$ and all their derivatives up to order $l - 1$, is a nonanticipating Markov extension of the original process $H(t)$, $-\infty < t < \infty$, because each value $\xi_k(t)$ is a linear combination of the components $\dot{\xi}_j(t)$ and their derivatives at t; the coefficients in this linear combination are elements of the matrix-coefficients of the polynomial $P(z)$ of degree $\leq l - 1$ in the original factorization formula for the spectral density $f(\lambda)$.

In short, with formula (3.6) in mind, we have the following result for a finite dimensional stationary process $H(t)$, $-\infty < t < \infty$, with matrix-valued spectral density $f(\lambda)$.

Theorem 3. *The minimal Markov extension*

$$P(H_{-\infty, t})H_{t, \infty}, \qquad -\infty < t < \infty,$$

is finite dimensional if and only if the spectral density $f(\lambda)$ is rational.

3. Stationary Processes with Symmetric Spectra

We turn to the "symmetric spectrum" case, assuming the existence of a spectral density $f(\lambda)$ such that

$$f(-\lambda) = f(\lambda), \qquad -\infty < \lambda < \infty; \tag{3.13}$$

we specify here that $f(\lambda)$ is a norm-integrable function whose values are positive operators on the Hilbert space X and the associated covariance function is

$$B(t) = \int_{-\infty}^{\infty} \exp(i\lambda t) f(\lambda)\, d\lambda, \qquad -\infty < t < \infty.$$

Under (3.13)

$$B^*(t) = B(-t) = \int_{-\infty}^{\infty} \exp(-i\lambda t) f(\lambda)\, d\lambda = \int_{-\infty}^{\infty} \exp(i\lambda t) f(\lambda)\, d\lambda = B(t),$$

so that in (3.10) we are dealing with a continuous semigroup of positive self-adjoint operators

$$B(t) = B^*(t/2)B(t/2), \qquad t \geq 0.$$

As we know, the exponential formula (3.11) is true for such a semigroup under the condition

$$\|B(t)\| < 1, \qquad t \neq 0,$$

and in this case

$$B(t) = \int_{0}^{\infty} \exp(-\alpha t) P(d\alpha), \qquad t \geq 0, \tag{3.14}$$

where $P(d\alpha)$, $\alpha > 0$, is the family of projections in the spectral representation

$$-a = -\int_0^\infty \alpha P(d\alpha)$$

of the infinitesimal generator

$$-a = \lim_{h \to 0} \frac{B(h) - I}{h}$$

(a is a positive self-adjoint operator).

As an illustration of our assertions we consider an example which is interesting in its own right.

Let

$$\xi(u), \qquad u \in C_0^\infty(\mathbb{R}^d)$$

be a stationary generalized random function with spectral density of the form

$$f(\lambda, \mu) = \frac{1}{a(\mu)^2 + \lambda^2}, \qquad \lambda \in \mathbb{R}^1, \mu \in \mathbb{R}^{d-1}, \qquad (3.15)$$

where $a(\mu)$ is a positive function of the variable $\mu \in \mathbb{R}^{d-1}$, with

$$0 < c_1 \le a(\mu)^2 \le c_2(1 + |\mu|^2)^l.$$

Let

$$X = H_+(\Gamma),$$

where Γ is the boundary separating the half-spaces

$$S_1 = (-\infty, 0) \times \mathbb{R}^{d-1}, \qquad S_2 = (0, \infty) \times \mathbb{R}^{d-1};$$

remember that

$$H_+(\Gamma) = \bigcap_{\varepsilon > 0} H(\Gamma^\varepsilon),$$

and $H(\Gamma^\varepsilon)$ is the closed linear span of the variables

$$\xi(u), \qquad \text{Supp } u \subseteq \Gamma^\varepsilon = (-\varepsilon, \varepsilon) \times \mathbb{R}^{d-1}.$$

In view of the isomorphism $\eta \leftrightarrow \varphi(\lambda, \mu)$, which is given by the spectral representation

$$\eta = \int_{\mathbb{R}^1} \int_{\mathbb{R}^{d-1}} \varphi(\lambda, \mu) \Phi(d\lambda \, d\mu),$$

see Chapter 3, (2.5), we can say that $H_+(\Gamma)$ is the space of all functions $\varphi(\mu)$, $\mu \in \mathbb{R}^{d-1}$, such that

$$\int_{\mathbb{R}^{d-1}} |\varphi(\mu)|^2 g(\mu) < \infty,$$

where

$$g(\mu) = \int_{\mathbb{R}^1} f(\lambda, \mu) \, d\lambda, \qquad \mu \in \mathbb{R}^{d-1}.$$

We will show this. For a spectral density of type (3.15) the associated duality condition is true with respect to all open domains $S \subseteq \mathbb{R}^d$ (cf. Chapter 3, §3) and for $S = S_1 \cup S_2$ we have

$$H_+(\Gamma)^\perp = H^*(S_1 \cup S_2).$$

It is clear that each function $\varphi(\lambda, \mu) \in H_+(\Gamma)$ is defined by the orthogonality condition

$$0 = \int_{\mathbb{R}^1} \int_{\mathbb{R}^{d-1}} [\tilde{u}(\lambda)\tilde{v}(\mu)]\gamma(\lambda, \mu)\overline{\varphi(\lambda, \mu)}f(\lambda, \mu) \, d\lambda \, d\mu$$

$$= \int_{\mathbb{R}^{d-1}} \tilde{v}(\mu)\left[\int_{\mathbb{R}^1} \tilde{u}(\lambda)\overline{\varphi(\lambda, \mu)} \, d\lambda\right] d\mu, \qquad v \in C_0^\infty(\mathbb{R}^{d-1}), \; u \in C_0^\infty(\mathbb{R}\backslash\{0\}),$$

where

$$\gamma(\lambda, \mu) = 1/(2\pi)^d f(\lambda, \mu)^{-1},$$

see Chapter 3, (2.16), and

$$\int_{\mathbb{R}^1} \tilde{u}(\lambda)\overline{\varphi(\lambda, \mu)} \, d\lambda = 0, \qquad \text{a.e. } \mu \in \mathbb{R}^{d-1}$$

for all $u \in C_0^\infty(\mathbb{R}\backslash\{0\})$. It is clear that as a function of $\lambda \in \mathbb{R}$, $\varphi(\lambda, \mu)$ is the Fourier transform of a generalized function with support $\{0\}$ and thus $\varphi(\lambda, \mu)$ is a polynomial in $\lambda \in \mathbb{R}$. Taking into account the condition

$$\int_{\mathbb{R}} |\varphi(\lambda, \mu)|^2 f(\lambda, \mu) \, d\lambda < \infty, \qquad \text{a.e. } \mu \in \mathbb{R}^{d-1},$$

where $f(\lambda, \mu)$ is of the form (3.15), we finally get that the function $\varphi(\lambda, \mu)$ is constant in the variable $\lambda \in \mathbb{R}$, $\varphi(\lambda, \mu) = \varphi(\mu)$, which is what we wanted to show.

Let us look at the stationary process

$$U_t X, \qquad -\infty < t < \infty, \qquad (3.16)$$

where $X = H_+(\Gamma)$ and $U_t = U_{(t, 0)}$ is the unitary group connected with the original stationary function: U_t is the shift along the coordinate axis of the variable $t \in \mathbb{R}$. As we have seen already, the space X can be identified with the standard Hilbert space of functions $\varphi(\mu)$, $\mu \in \mathbb{R}^{d-1}$, which are square-integrable with respect to the weighting $g(\mu)$. It is clear that our stationary process has a spectral density $f(\lambda)$ which can be represented as the operation of multiplication by the function

$$\frac{f(\lambda, \mu)}{g(\mu)}, \qquad \mu \in \mathbb{R}^{d-1}; \qquad (3.17)$$

this follows from the equation

$$\int_{\mathbb{R}} \exp(i\lambda t) \int_{\mathbb{R}^{d-1}} [f(\lambda)\varphi(\mu)]\overline{\psi(\mu)}g(\mu) \, d\mu \, d\lambda$$

$$= \int_{\mathbb{R}} \int_{\mathbb{R}^{d-1}} \exp(i\lambda t)\varphi(\mu)\overline{\psi(\mu)}f(\lambda, \mu) \, d\lambda \, d\mu$$

for any $\varphi, \psi \in X$. It is easy to check that *the covariance function*

$$B(t) = \int_{-\infty}^{\infty} \exp(i\lambda t)f(\lambda) \, d\lambda$$

on the Hilbert space X is the operator multiplication by the function

$$\exp[-a(\mu)|t|], \qquad \mu \in \mathbb{R}^{d-1}, \tag{3.18}$$

and we are dealing with the nondegenerate semigroup of positive operators $B(t)$, $t \geq 0$, whose infinitesimal generator is the operator multiplication by $-a(\mu)$, $\mu \in \mathbb{R}^{d-1}$.

Note that the Markov behavior of the stationary process (3.16) implies the Markov property for the stationary function with spectral density (3.15) with respect to the half-spaces

$$S = (-\infty, t) \times \mathbb{R}^{d-1} \subseteq \mathbb{R}^d.$$

Notes

Chapter 1

§1. A more detailed exposition of the material found here can be found in various texts on measure theory and probability theory; in particular, a systematic presentation of questions in distribution theory on metric spaces is found in [41].

§2–4. The choice of conditional distributions described here is used in the theory of dynamic systems for the decomposition into ergodic components (in this connection see [11], for example). The general question of the existence of consistent conditional distributions is considered here for the first time, apparently. The related results in §4.3 were obtained (and presented here) by S. E. Kuznetzov. Consistent conditional distributions for discrete "random fields" have found important application in certain problems in statistical physics (the problems involved and related literature can be found in [42], for example).

§5. Various uses of Hermite polynomials in Gaussian distributions are well known; the properties of Hermite polynomials of several variables described here—in particular, the behavior of their conditional expectations—were noted previously in [13]. Multiple stochastic integrals were introduced into probability theory in [28].

Chapter 2

§1. One of the first examples of the traditional Markov property being treated in a wider sense is provided by so-called complex Markov chains. Another example is offered by the class of processes whose behavior outside a finite time interval, given fixed values on its endpoints, does not depend on their behavior inside this interval (Gaussian processes of this type were considered in detail in [29]; earlier, an analogous "pure Markov property" was considered for Gaussian isotropic fields in [23], [49]). It seems that the first formulation of a generalized Markov random process was presented in [48]. One of the first investigations of the Markov property for random functions of several variables was carried out for Levy–Brownian motion in [35], introducing the idea of a splitting σ-algebra.

The term "random field" has been applied more than once, with various meanings, reflecting the presence of a multi-dimensional parameter space T; mostly it has been used in studying random functions $\xi(t), t \in T$, but a wider significance was given to the term in a series of works investigating σ-algebras of events $\mathscr{A}(S), S \subseteq T$, which satisfy the monotonicity condition (1.25)—see, for example, [5], [6], [3]. Our idea of a random field $\mathscr{A}(S), S \subseteq T$, includes the additivity condition (1.24), without which many important propositions about the Markov property could not be proved or even would be false. We have in mind here, before all else, the fact that with the additivity condition the extension of the appropriate splitting σ-algebra, necessary in many cases, again gives us a splitting σ-algebra. Moreover, additivity turns out to be the property whose presence allows us to give convenient criteria for Markov behavior in terms of the conjugate random field, cf. §3.

The Markov property, as applied to a random field $\mathscr{A}(S), S \subseteq T$, usually has been thought of, until recently, as meaning that for mutually disjoint domains $S_1 = S$ and $S_2 = T \backslash \bar{S}$ with boundary $\Gamma = \partial S$, the boundary σ-algebra $\mathscr{A}_+(\Gamma)$ splits $\mathscr{A}_+(S_1)$ and $\mathscr{A}_+(S_2)$, see (1.29); in other words, taking advantage of the terminology of random processes we could say that for a given "phase state" $\mathscr{A}_+(\Gamma)$, the "future" $\mathscr{A}_+(S_2)$ is independent of the "past" $\mathscr{A}_+(S_1)$. But for such a Markov property the extension $\mathscr{A}(\Gamma^\varepsilon)$ of the σ-algebra $\mathscr{A}_+(\Gamma)$, which is necessary in many cases, may not even be a "phase state"; more precisely, $\mathscr{A}(\Gamma^\varepsilon)$ may not even be splitting for $\mathscr{A}(S_1)$ and $\mathscr{A}(S_2)$. This appears to be typical for random fields generated by generalized random functions. Under our somewhat stronger Markov property (see (1.27)) any extension of the "phase state" of the form $\mathscr{A}(\Gamma^\varepsilon)$ again gives us a "phase state," given which, the "future" $\mathscr{A}(S_2)$ does not depend on the "past" $\mathscr{A}(S_1)$.

§2. Stopping σ-algebras in relation to the strong Markov property for random processes were investigated in [26], [36]. Our results were largely inspired by [3].

§3. Markov properties of type (3.14) for Gaussian random fields, which were the starting point for our investigations, have appeared in many works. In some of these ([5], [30]), conditions (3.14) are rephrased in terms of a "conjugate field" defined by (3.24) and, generally speaking, not having the additivity property; without this present in the associated Markov conditions, it seems very difficult to study the first of conditions (3.14) (more accurately, the first of conditions (3.10) or (3.12)). Our conjugate field is additive by definition and its introduction is very convenient from a methodological point of view, essentially allowing us to reduce the study of the Markov property to a question of the existence of an orthogonal conjugate field.

Chapter 3

§1. Much information about generalized random functions can be found in [1].

The idea of biorthogonal generalized functions in connection with the Markov property was introduced in [5]. The inclusion (1.13), which is equivalent to the existence of a biorthogonal function, was used to study the Markov property in [43]. The duality conditions (1.21), (1.23) were derived in [17], [46].

§2. The basic statements of the spectral theory of stationary functions can be found in [1] and [11]. The theorem about Markov stationary functions whose spectral densities have the form $f(\lambda) = 1/g(\lambda)$, where $g(\lambda)$ is a polynomial in $\lambda \in \mathbb{R}^d$, was formulated in [5]; under the additional restriction

$$f(\lambda) \sim (1 + |\lambda|^2)^{-l},$$

where $l > d/2$ is an integer, a similar result appeared at the same time in [43].

§3. The duality condition (3.12) and its consequences for the Markov property were obtained in [17]. Elliptic bilinear forms of the type (3.14) were used in connection with the Markov property in [43].

§4. Levy–Brownian motion was introduced in [34]; its Markov behavior was shown in [35] with the use of "white noise," introduced in [22]; other properties were studied in [4].

The basic propositions about the properties of spaces of the form $\mathring{W}_2^l$ can be found in [8], [18]. The continuous renewal of a Markov field (Theorem 1) was noted in [17]. The structure of the space $H_+(0)$ for a stationary process has been used in many works (see, for example, [33], [19]). A statement about the structure of the boundary space $H_+(\Gamma)$, similar to Theorem 2 but for the case of an elliptic form of order $2l > d$, where d is the dimension of the parameter space, is found in [43]; Theorem 2 was obtained in [17]. The Markov property, given a correlation function satisfying the generalized equation analogous to (4.25), was mentioned in [7]. The Dirichlet problem for nonzero boundary conditions (4.30) was studied in [46].

§5. A linear stochastic differential equation with partial derivatives of the form (5.15) was investigated in connection with the Markov property in [17], cf. also [50]. The problems of interpolation and extrapolation were considered for various classes of random functions in [4], [17], [43]. The "Brownian sheet," for a parameter space of any dimension, introduced in [21] under the name "Wiener field" as an example of the Markov random function, was considered in [17] (see also [50]); among the possible applications of this function we mention the stochastic integral on the plane, which uses it in the definition (cf. [24], for example).

Chapter 4

§1. Basic information about vector-valued stationary functions can be found in [11], [14]. Conditions (1.12)–(1.14), with which we showed the existence of a conjugate field, were used in [16]; similar conditions were mentioned implicitly in the short note [6] in looking at the Markov property for scalar stationary functions. A condition of type (1.16) appeared earlier in connection with the investigation of the analytic structure of the space $H(S)$ in [9], [10]; a similar condition was used in [32] to study the Markov property of scalar stationary functions.

§2. The Markov property for vector-valued stationary function was studied in [16], [46]. The idea of an L-Markov random field, which reflects the specific character of a splitting boundary of type (2.1), was introduced for a discrete parameter t in [12]; it corresponds better with the possibilities of spectral methods (the symbol L was used to denote the neighborhood from which one constructs the boundary separating the "past" from the "future"). The L-Markov property for continuous scalar stationary fields was considered in [6], [45], where L-Markov conditions similar to (2.7) were presented. The Markov property for a scalar stationary function having spectral density $f(\lambda) = 1/g(\lambda)$, where $g(\lambda)$ is an entire function of minimal exponential type, was established in [32].

§3. The material contained here is approximately the same as in [15]. The example of the stationary process (3.16) was presented in connection with work in [37].

Bibliography

[1] Gel'fand, I. M. and Vilenkin, N. J. *Generalized Functions*, vol. 4. *Some Applications of Harmonic Analysis*. Fizmatgiz: Moscow, 1961. English transl., Academic Press: New York, 1964.

[2] Dobrushin, R. L. and Minlos, R. A. Polynomials in linear random functions. *Uspekhi Matem. Nauk*, **32**, No. 2 (1971). English transl. in *Russian Math. Surveys*, **32**, no. 2 (1971), 71–127.

[3] Evstigneev, I. V. Markov times for random fields. *Teor. Verojatnost. i Primenen.*, **22**, no. 3 (1977), 575–581. English transl. in *Theor. Probability Appl.*, **22**, no. 3 (1977).

[4] Molchan, G. M. Some problems for Lévy's Brownian motion. *Theor. Probability Appl.*, **12**, no. 3 (1967), 682–690.

[5] Molchan, G. M. Characterization of Gaussian fields with the Markov property. *Dokl. Akad. Nauk SSSR*, **197**, (1971), 784–787. English transl. in *Soviet Math. Dokl.*, **12** (1971).

[6] Molchan, G. M. L-Markov Gaussian fields. *Dokl. Akad. Nauk SSSR*, **215** (1974), 1054–1057. English transl. in *Soviet Math. Dokl.*, **15** (1974).

[7] Molchan, G. M. The Markov property of Lévy fields on spaces of constant curvature. *Dokl. Akad. Nauk SSSR*, **221**, no. 6 (1975), 1276–1279. English transl. in *Soviet Math. Dokl.* **16-1**, no. 2 (1975), 528–532.

[8] Nikolsky, S. M. *Approximation of Functions of Several Variables and Embedding Theorems*. Nauka: Moscow, 1969. English transl., Springer-Verlag: Berlin and New York, 1974.

[9] Presnjakova, O. A. On the analytic structure of subspaces generated by homogeneous random fields, *Dokl. Akad. Nauk SSSR*, **192** (1970), 279–281. English transl. in *Soviet Math. Dokl.*, **11** (1970).

[10] Orekova, O. A. Some problems of extrapolation of random fields. *Dokl. Akad. Nauk SSSR*, **196** (1971), 776–778. English transl. in *Soviet Math. Dokl.*, **12** (1971).

[11] Rozanov, Ju. A. Stationary Random Processes. Fizmatgiz: Moscow, 1963. English transl., Holden-Day: San Francisco, Calif., 1967.

[12] Rozanov, Ju. A. Gaussian fields with given conditional distributions. *Teor. Verojatnost. i Primenen*, **12** (1967), 433–443. English transl. in *Theor. Probability Appl.*, **12** (1967).

[13]   Rozanov, Ju. A. Infinite-dimensional Gaussian distributions. *Proceedings of the Steklov Inst. of Mathematics, no. 108 (1971)*. Amer. Math. Soc.: Providence, R.I., 1971.

[14]   Rozanov, Ju. A. *Theory of Renewal Processes*. Nauka: Moscow, 1974. (Russian).

[15]   Rozanov, Ju. A. Markov extensions of a random process. *Teor. Verojatnost. i Primenen.*, **22**, no. 1 (1977), 195–199. English transl. in *Theor. Probability Appl.*, **22**, no. 1 (1977).

[16]   Rozanov, Ju. A. On the theory of random stationary fields. *Mat. Sb.* **103**(145), No. 1(5) (1977) 3–22.

[17]   Rozanov, Ju. A., On Markovian fields and stochastic equations. *Mat. Sb.* **106**(148), No. 1(5), (1978), 106–116.

[18]   Sobolev, S. L. *Applications of Functional Analysis in Mathematical Physics*. Izdat. Leningrad. Gos. Univ.: Leningrad, 1950. English transl., Amer. Math. Soc.: Providence, R.I., 1963.

[19]   Tutubalin, V. N. and Freidlin, M. I. The structure of the infinitesimal σ-algebra of a Gaussian process. *Teor. Verojatnost. i Primenen.*, **7** (1962), 204–208. English transl. in *Theor. Probability Appl.*, **7** (1962).

[20]   Chiang Tse-pei. Extrapolation theory of a homogeneous random field with continuous parameters. *Teor. Verojatnost. i Primenen.* **2** (1957), 60–91. English transl. in *Theor. Probability Appl.*, **2** (1957).

[21]   Chentsov, N. N. (Cencov). Wiener random fields with several parameters. *Dokl. Akad. Nauk SSSR*, **106**, no. 4 (1956), 607–609.

[22]   Chentsov, N. N. Lévy-type Brownian motion for several parameters and generalized white noise. *Teor. Verojatnost. i Primenen.*, **2** (1957), 281–282. English transl. in *Theor. Probability Appl.*, **2** (1957).

[23]   Jadrenko, M. I. Isotropic Gaussian random fields of Markov type on a sphere. *Dopovidi Akad. Nauk Ukrain. RSR*, **1959**, 231–236. (Ukrainian)

[24]   Carroli, P. and Walsh, J. B. Stochastic integrals in the plane. *Acta Math.*, **134**, nos. 1–2 (1975), 11–183.

[25]   Cartier, P. Introduction a l'etude des mouvements Browiens a plusiers parametetres. *Seminaire de Probabilities V*. Lecture Notes in Mathematics, vol. 191, pp. 58–75. Springer-Verlag: New York, 1971.

[26]   Chung, K. L. and Doob, J. L. Fields, optionality and measurability. *Amer. J. Math.*, **87** (1965), 397–424.

[27]   Hida, T. *Stationary Stochastic Processes*. Princeton, N.J.: Princeton University Press, 1970.

[28]   Ito, K. Multiple Wiener integral. *J. Math. Soc. Japan*, **3** (1951), 157–169.

[29]   Jamison, B. Reciprocal process. *Z. Wahr. verw. Geb.*, **30** (1974), 65–86.

[30]   Kallianpur, G. and Mandrekar, U. The Markov property for generalized Gaussian random fields. *Ann. Inst. Fourier*, **24**, no. 2 (1974), 143–167.

[31]   Knight, F. A remark on Markovian germ fields. *Z. Wahr. verw. Geb.*, **15** (1970), p. 291–296.

[32]   Kotani, S. On a Markov property for stationary Gaussian processes with a multi-dimensional parameter. *Proceedings of the Second Japan–USSR Symposium on Probability*, Lecture Notes in Mathematics, vol. 330, pp. 239–250. Springer-Verlag: New York, 1973.

[33]   Levinson, N. and McKean, H. I. Weighted trigonometrical approximation with applications to the germ field of a stationary Gaussian noise. *Acta Math.*, **112** (1964), 99–143.

[34]   Levy, P. A special problem of Brownian motion, and a general theory of gaussian random functions. *Proceedings of the Third Berkeley Symposium on Mathematical and Statistical Probability, 1956*, vol. 2, pp. 133–175.

[35]   McKean, Jr., H. P. Brownian motion with a several-dimensional time. *Teor. Verojatnost. i Primenen.* **8** (1963), 357–378.

[36] Meyer, P. A. and Dellacherie, C. *Probabilities and Potential.* North-Holland: Amsterdam, 1978.

[37] Nelson, E. Construction of quantum fields from Markoff fields. *J. Funct. Anal.,* **12**, no. 1 (1973), 97–112.

[38] Okabe, Y. Stationary Gaussian processes with Markovian property and M. Sato's hyperfunctions. *Japan J. Math.,* **41** (1973), 69–122.

[39] Okabe, Y. On stationary linear processes with Markovian property. *Proceedings of the Third Japan-USSR Symposium on Probability,* Lecture Notes in Mathematics, vol. 550, pp. 461–466. Springer-Verlag: New York, 1976.

[40] Okabe, Y. On the germ fields of stationary Gaussian processes with Markovian property. *J. Math. Soc. Japan,* **28** (1976), 86–95.

[41] Parthasarathy, K. *Probability Measures on Metric Spaces.* Academic Press: New York, 1967.

[42] Preston, C. *Random Fields.* Lecture Notes in Mathematics, vol. 34. Springer-Verlag, Berlin, 1976.

[43] Pitt, L. D. A Markov property for Gaussian processes with a multi-dimensional parameter. *Arch. Rat. Mech. Anal.,* **43** (1971), 367–391.

[44] Pitt, L. D. Some problems in the spectral theory of stationary processes on $\mathbb{R}^d$. *Indian Math. J.,* **23**, no. 4 (1973), 343–366.

[45] Pitt, L. D. Stationary Gaussian–Markov fields on $\mathbb{R}^d$ with deterministic component. *J. Multivariate Analysis,* **5**, no. 3 (1975), 300–313.

[46] Rozanov, Yu. A. Random Markov fields. In *Recent Developments in Statistics.* Academic Press: New York, 1979.

[47] Surgailis, D. On the Markov property of a class of linear infinitely divisible fields. *Z. Wahr. verw. Geb.,* (1979).

[48] Urbanik, K. Generalized stationary processes of Markovian character. *Studia Math.,* no. 3 (1962), 261–282.

[49] Wong, E. Homogeneous Gauss–Markov fields. *Ann. Math. Stat.,* **40**, no. 5 (1969), 1625–1634.

[50] Zerakidze, Z. S. On the Markov property of the solution of the hyperbola type differential equation. *Twelfth European Meeting of Statisticians, Varna, September 3-7, 1979,* Abstracts, p. 250.

Index

www.ingramcontent.com/pod-product-compliance
Lightning Source LLC
Chambersburg PA
CBHW070031280425
25805CB00010B/976